MW00479942

Anne Crawford was a feature writer for *The Age* and *The Sunday Age* for many years. She has co-authored three memoirs: *Shadow of a Girl* (Penguin, 1995), *Doctor Hugh, My Life with Animals* (Allen & Unwin, 2012), and the award-winning *Forged with Flames* (Wild Dingo Press, 2013).

Anne is also a published and exhibited photographer. She researched a documentary on South Africa, and worked as a volunteer in Nepal for the Fred Hollows Foundation, contributing to *Through Other Eyes* (Pan Macmillan, 2002). A born-again horse rider, she now owns a small mare with a lot of attitude called Poppy. Anne lives by the sea in rural South Gippsland.

Great Australian Horse Stories

Published by Allen & Unwin in 2013

Copyright © Anne Crawford 2013

'The Barkly Brute' song lyrics on page 64 reproduced with permission from the Estate of Stan Coster.

All rights reserved. No part of this book may be reproduced or transmitted in any form or by any means, electronic or mechanical, including photocopying, recording or by any information storage and retrieval system, without prior permission in writing from the publisher. The Australian *Copyright Act 1968* (the Act) allows a maximum of one chapter or 10 per cent of this book, whichever is the greater, to be photocopied by any educational institution for its educational purposes provided that the educational institution (or body that administers it) has given a remuneration notice to Copyright Agency Limited (CAL) under the Act.

Allen & Unwin
83 Alexander Street
Crows Nest NSW 2065
Australia
Phone: (61 2) 8425 0100
Fax: (61 2) 9906 2218
Email: info@allenandunwin.com
Web: www.allenandunwin.com

Cataloguing-in-Publication details are available
from the National Library of Australia
www.trove.nla.gov.au

ISBN 978 1 74331 680 1

Set in 12/16 pt Goudy Oldstyle Std by Bookhouse, Sydney
Printed and bound in Australia by Griffin Press

10 9

The paper in this book is FSC® certified.
FSC® promotes environmentally responsible,
socially beneficial and economically viable
management of the world's forests.

To my wonderful family

To my dear friend Rodders

CONTENTS

PART I *Wild horses*

1	Bogong Jack	3
2	A pony called Parrot	14
3	The man with the horse in his car	20
4	Saving the west's wild horses	25
5	Rumpy and the last run	35

PART II *Droving days and outback ways*

6	Rocket, an outback legend	41
7	The business of droving	47
8	Night horses revealed	56
9	The breaker, the brute and the queen	61
10	Mustering on	67

PART III *Time of their lives*

11	The man who soared high	75
12	A boyhood of horses	84
13	The accidental drover	94

PART IV *Riding the highway of life*

14 Journey of a lifetime 105
15 They call him Tex 114
16 Trotting back from the brink 121

PART V *High achievers*

17 Harry's dream 135
18 Made in Australia: The world's tallest horse 142
19 The horse that made the man 155

PART VI *Horses that help*

20 Sally's healing horses 169
21 To the rescue 177
22 A different approach 183

PART VII *Tales of the unexpected*

23 The horse on the road 193
24 Surprise moments for an equine vet 197
25 Beware the free horse 201

PART VIII *Hanging on to our heritage*

26 The *real* Geebung Polo Club 209
27 Holding on to the heavy horses 220
28 Thrill and spills of harness 228
29 Trooper: A horse given a fighting chance 233

PART IX *Beating the odds*

30 Brave Anne meets her match 247
31 Clancy 264

PART X *There's always one*

32	Beth and Gypsy	273
33	A horse called Spook	279
34	Zelie Bullen and Bullet	285
35	The Pact	294

Glossary 300
References 307
Acknowledgements 309

PART I

Wild horses

1

BOGONG JACK

Craig Orchard remembers the first time he saw Bogong Jack in the Victorian high plains with the clarity of air after rain. The black-and-white stallion, bigger and flashier than most, was standing with a mob of brumbies in a clearing, a grassy plain rimmed by woodland. The horse turned to look at him with his half-white face and snorted. The brumby runner from Benambra way had caught plenty of wild horses before but this one looked pretty special.

Craig reckons he first crossed paths with the stallion in the early 90s, before the bush was burnt in the 2003 fires and before he married Tahnee. Craig was taking cattle to the Bogong High Plains for an old cattleman called Charlie McNamara, who had a lease up there then. It was the middle of summer and the horses were taut with condition and full of energy.

But Bogong Jack, as he became known, wasn't having a bar of the horse and rider that appeared on his patch. The stallion flicked his tail in the air, wheeled around towards his mares and galloped off with the rest of the herd into the scrub. Craig took off, chasing him, but he got away. The high plains are hard country to pursue horses: steep, rocky, full of rabbit and wombat holes, logs and little mongrel creeks with banks that subside under hoof. But he'd catch the stallion next time, he swore to himself. He never forgot a horse once he'd seen it. He'd be back for him.

Craig had been fascinated by the wild horses ever since he was a boy, sitting round the campfires listening to the 'old fellas' talk about brumby running. He helped a cattleman from Bindi called Mick Murphy drive his herefords—200 of them—up to the high plains every year and there were usually brumbies on the way and stories about them at night; tales of men trying to yard a mob, the spectacular ways they got caught, the cunning brumby that always escaped. Stories that stretched with the telling like shadows in a setting sun.

The brumbies were kept in a holding paddock until there were enough of them to make it worth bringing them down to be trucked to Omeo. Most of them were sold to be broken in and ridden by the children of cattlemen or farmers, though Charlie McNamara used to take a few down to the Omeo Rodeo. Didn't get their nickname of 'bucks' for nothing.

Craig got his first catching rope at the age of eleven—not a fancy plaited cowhide like the older men and rodeo riders used, but one he made on the farm from baling twine twisted into rope with a winder. As he made it he thought to himself, 'I'm going to catch one of these brumbies.'

But it wasn't that easy. The first time he went after a brumby the horse outran him and Craig had to pull back, not wanting to go too far after it and get lost. Took him a couple of years before he learnt his way round the bush enough to rope one, but after that he got a bit faster and went further each time. His confidence grew, and from about the age of fifteen he was catching big numbers.

He started brumby running at seventeen with Jock Sievers, a rugged bushman who worked as a logger with Craig's father and wanted someone to catch horses with him. In time, Jock's daughters Tahnee and Aleshia came along with them. There's a fair few girls that do brumby running and plenty of them are as good as the blokes, Craig says. Tahnee was handy on a horse and became a fine whip-cracker. A good looker, too, with that smile and her long dark hair.

Craig would go out with the Sievers and others through from Limestone Creek out of Benambra to Quambat at the head of the Murray River to Nunniong back towards Swifts Creek, the Bogongs beside Mount Hotham and Spring Creek. Wherever there were brumbies, they'd go.

There's not much he doesn't know about brumby catching now. By the time he was 44 years old in 2013, he'd notched up more than 900 catches. It's a real good sport, he says, it's you against them. Once it gets in your blood you can't stop.

Craig has a licence to take brumbies out of the state forests as a subcontractor for Parks Victoria, or 'Parks' as it's called. He became part of what started as the Australian Brumby Management Association about twenty years ago, an organisation that made sure that people licensed to catch brumbies treated them humanely. The contractors remove a set number of horses every year from the national park as part of a program to keep numbers down. Craig tries to catch the younger ones—from eight months to two years old—because they're easier to find homes for. The older horses aren't good for much: hard to train and hard to fence in. He and the other contractors aren't allowed to release horses they catch back in the bush so if they catch a horse that turns out to be old, its fate is usually sealed. Unless someone wants it—and most people don't want an old brumby—it will be sent to the saleyards or knackery. He'd rather not catch the old stallions or mares but says he knows that if the number of brumbies in the parks isn't controlled, it could be worse for the horses. 'I'd hate to see them all get shot out,' he says.

In general, the brumbies see you then take off flat-out one way and you take off after them exactly where they go and just keep going and going until one 'knocks up' or you get faster than them and head them off. If you can't catch one in the first 200 or 300 metres, galloping up alongside them and looping a rope around their neck, you might have to keep chasing them for 3 or 4 kilometres. Craig once rode after a brumby for more than 13 kilometres. It was near Nunniong on the logging roads and he retraced the chase in the ute, checking how far he'd gone on the speedometer.

He got that one but there is always the one that gets away; the wily brumby that isn't going to be outwitted by a man on a thoroughbred. Like the baldy-faced mare at Spring Creek at the back of Cobungra Station, the state's biggest cattle station. An old story went that years ago the McCrae family, who had a bush hut there, released a Clydesdale stallion that was no longer of use to them. Cattlemen and farmers did this from time to time to improve the look of the brumbies, sometimes inbred and ugly as sin. The occasional trotter would go bush, or a coloured stallion, just to make the breeding better. The McCrae's Clydie mixed in with the local brumby mares and bred up some good-looking horses: brumbies with four white feet, chestnuts and blacks with big white faces.

At the time he saw the baldy-faced mare, Craig was competing in a few mountain races and had a fast horse, a big old thoroughbred that had been a racehorse when it was young and was now surefooted in the bush. Craig and his mates found a mob of brumbies in an area where the bush opened out to a meadow. Craig was winding up the thoroughbred, approaching the moving mob, thinking he was doing a 'real good job' of keeping up with them. He noticed the big mare with a lot of Clydesdale in her and thought she'd make a good broodmare. He galloped past a big foal and a yearling heading towards her, thinking he'd slip in and get her 'easy', sure that she couldn't outrun him on the thoroughbred. He drew in beside the mare and, moving quickly, got the rope ready to drop round her neck. The mare looked at him sideways, pulled away, stretched her neck and left him for dead.

Craig laughs and shakes his head thinking about it. 'I thought, I'm on a racehorse in an open gallop and she pulled away. It was like chasing a Melbourne Cup horse! By geez she could go. I admire them sort of horses.' He dropped the rope on the yearling and took him home instead. The mare's probably still out there or died of old age, Craig says. And good on her.

Then there are the rogue horses, the horses that take you on rather than galloping away from you.

There were a couple of pretty bad stallions up the back of Nunniong; horses that would see you and come at you flat-out to protect their mares. One day Craig was with an old cattleman from Swifts Creek and a good mate, Dean, riding on the Mia Mia logging road. The men had just started a mob off when Craig galloped into the herd, heading for a mare he wanted to catch. He looked round at what he thought was Dean approaching, and saw a set of teeth coming at him instead. A big black stallion, ears flattened, necked strained, lips back baring his teeth and lunging at him. Craig gathered the catching rope coiled on his horse's neck and smacked the stallion over the head with it as he passed to ward it off. The stallion dropped in behind him and Craig moved closer to the mare.

He'd chased her for a kilometre down the road when he heard a horse coming up behind him. 'Get off me horse,' he yelled at what he thought was Dean getting too close to the thoroughbred's rump. 'Get off me horse!'

Craig was leaning down about to loop the rope round the head of the mare when the black stallion's head appeared again from around the side, coming in to grab him by the thigh. Ripped him clean off his horse at a gallop, rolled him down the logging road, splitting his knee on the rocks and gravel. The stallion peeled away and off. Thought he'd won.

Dean flew past pursuing the mob and had been going for about a kilometre when the black stallion caught up and got hold of him, biting him hard on the shoulder. The two men met up and swapped accounts. Craig went back to have his leg stitched.

Quite a few people got bluffed by that horse. Big black thing, horror of a horse that didn't like riders, Craig recalls. Doesn't think he ever got caught.

There was another stallion that gave them hell, a small bay horse that ran with a huge mob of mares, yearlings and foals near what the locals know as 'Big Bend' on the Livingstone River. You were in for a fight as soon as you saw that horse, you knew you'd get a doing. The stallion would come at horse and rider, grab onto the horse by the neck and hang on fit to rip it. 'Geez, don't get me leg!' Craig would swear under his breath. He and Dean would split up one riding above the stallion on the slope, the other below the wily little horse. But the stallion would zigzag between the riders, evading them, then latch onto the nearest horse and try to tear its neck.

After ten or fifteen runs, Craig cracked it. 'I said to me mate this day, "I've had a gutful of this horse, I'm going to just gallop in at him flat-out and ram him. See how you go trying to get round him."' He headed his horse towards the mob. The stallion turned and in true form started to barrel down on them. Craig galloped as hard as he could into the horse and cleaned him up, knocking him to the ground. He looked down. His horse was standing right on top of the brumby—he had him! The stallion rolled his eyes and thrashed his head and legs around. 'I'm going to rope this bugger,' Craig said as he dipped the loop round the stallion's twisting, turning head. But he kept missing as the stallion kicked out and struggled below him. Kept swiping the rope round his neck like a madman. The stallion managed to crawl out, struggle to his feet and took off back to his mares. Craig watched his departing rump and shook his head. A real bad-natured horse, really hated people, too, he recalls. 'I reckon he would've taken to you if you were standing with a fishing line along the river, that fella.'

Dean caught the stallion the next time the pair went out after him and Craig came back with another horse out of the mob. But it was a real shame to take that stallion out of the bush, he says, with admiration.

Bogong Jack continued to elude Craig for months in the high plains. Craig came across the black-and-white brumby maybe a dozen times, sending his cattle dogs into the bush hoping to flush him out of the scrub, chasing him whenever he could. Summer turned into spring, turned into winter, and still the stallion kept his distance. Then, one day when Craig was riding through the snow in 'open tops' country—the big clearings in the high plains—he saw him again. The snow was a metre deep. Like a beautiful smooth white carpet. This time the wild horse couldn't slip away—he was leaving footprints everywhere.

Craig kept following the stallion and his tracks. But the thoroughbred had the advantage: his legs were longer and he moved through the snow more easily. They kept on the brumby's trail until they wore him out. For the last few hours Craig trotted, cantered and galloped after the stallion until he got him to what he thought was the right spot, then ran him down, galloping up behind and roping him. The brumby kicked the hell out of his horse. They get cranky protecting their mares and foals, Craig says, and will call out to them for a while afterwards. It was a big day—Craig roped a samba deer, too. He rode home carrying the deer's head on his horse and leading Bogong Jack on a halter.

It was around the time he caught the stallion in the early 2000s that he was talking to Jock Sievers and Tahnee about a dream Tahnee had. She wanted to create a brumby park, private land where brumbies could roam, safe from poachers and starvation, and where anyone who wanted to could see them in their natural state. They knew that people were fascinated by Australia's wild horse—the heritage and history that went with them, the spirit that inspired

Banjo Paterson to write about them in poems like *The Man from Snowy River* and Elynne Mitchell to celebrate them in her *Silver Brumby* books. But people rarely got a chance to see them. Jock had 500 acres of mostly bush that would do the job. They could run a number of brumbies in the park, which the public could be taken in to see, and keep training and rehoming them.

Craig had been helping Jock and Tahnee with the horses but was busy with farming so left the park plans to them. He and Tahnee were also due to get married in late 2003. The bushfires went through the area early that year, though, and the ceremony was delayed. The 2003 bushfires killed a lot more brumbies than the management program ever removed: hundreds of them. Craig was paid by a wildlife group to humanely destroy animals after the fires, riding through the charred area from Benambra to Mount Kosciusko. He found horses with no hair left on them but which were still alive, tails burnt off, hooves gone off their feet but still hobbling around. It wasn't just horses, either; there were deer, wallabies, wombats, kangaroos and all kinds of wildlife. He shot a handful of horses. There would've been a lot more he didn't find.

When Craig and Tahnee did get married, it was a pretty flash wedding. Held in February 2004 at Tahnee's grandfather's property in marquees near the banks of Limestone Creek. As the large crowd waited, Craig and three men rode in on horses one way, and Jock rode in from the other way with Tahnee in a big white wedding dress, riding sidesaddle, 'all made up'. The bridesmaids arrived in the back of a ute.

It was also around this time that a film producer who Craig had met years before reappeared. The producer had talked to him about a documentary series he wanted to make about brumbies, following a mob for a year. The crew spoke to Craig and the Sievers, heard them talk about their dreams for the brumby park, and the wedding, and changed their minds; they wanted to make the series about *them.*

'So we had to get married again,' Craig says, laughing. No guests this time and not as big but the re-enactment was pretty close. The three-part series called *Wild Valley* took the best part of a year to film. Craig and Tahnee and other riders were filmed, sometimes close-up from a helicopter, chasing brumbies, and building the cabin they were to live in for a while.

Now they live out of Benambra on a cattle farm with a couple of kids, Dallas and Bonnie. The brumbies Craig catches come home with him to be broken in and trained, and are sold or given away. They make great kids' ponies. They're easier to break in than a 'bred' horse such as a thoroughbred and are more docile. Once you show a brumby something, it sticks. Take that little black one that Aleshia broke in when she was twelve. They gave Parrot to a family in Sydney, where he's winning at shows. The family sent a photo of Parrot a while back, ribbons running the whole way up his neck.

They've got a paddock full of brumbies now, Craig says, shaking his head and laughing. Better than sending them to the knackery, where they end up as pet food. Yeah, he's got a bit of a soft spot for them. They do well in the paddocks once they're wormed, are free of lice, and fed. Some of the brumbies in the wild are pretty skinny, particularly during drought years followed by snow, and after fires. There are 80 to 100 brumbies at the Australian Alpine Brumby Park which Tahnee and Jock set up on his land in 2005.

Tahnee doesn't ride anymore. She's got a few problems now she's got motor neurone disease, that's sort of wobbled her up a bit, Craig says. She handles the sales of the brumbies or 'adopts' them out as part of the park. Gets a lot of calls about them.

Craig does less brumby running than he used to when he was younger and says he'll hang up the catching rope and retire at a thousand brumbies. Well, maybe.

Brumby running takes its toll on your body over the years. In the days before one of his mates discovered motocross knee pads, the riders' boots would be sticky with blood at the end of the day from running their shins into trees all day. The riding horses are skilled in the bush but you can't avoid every tree, Craig says. A cornered brumby will lash out at your horse and kick you in the shins doing it.

But Craig has been lucky with injuries. He's only had a couple of big ones. He dislocated his shoulder once, taking a tumble after his horse cartwheeled on a soft bank and landed on him. Shoulder's been buggered ever since, he says.

Broke his leg in several places once, too. He'd been training his old grey thoroughbred for the Tasmanian Cattlemen's Cup in 1991 and the horse jumped sideways into a post as he was going through a gateway a bit fast. Shattered his leg. Craig got off the horse to close the gate then realised he couldn't stand up; his leg was bending the wrong way. He couldn't get back onto the horse, either. At the time he was near the road to Benambra, a fair way from where he was staying with friends. He put the reins over his horse's neck and tried to shoo it away, thinking that if it went home fast without a rider they'd come looking for him. The horse just stood there staring at him—all his horses are trained to stand on the spot when he dismounts so that they don't move away when he gets off to work with cattle. Didn't matter what he did, the horse wouldn't leave him.

He tried waving down the drivers of some passing cars; they all waved back. Until finally a motorist who saw him struggle to get up and fall over stopped the car. They took him to his friends' house where Craig rang the doctor, who was at home at the time. Craig told him he'd smashed up his leg and would be coming in to the surgery. The doctor, who took a dim view of horsemen who pushed the limits, told him he was pruning the garden and he'd have to wait 45 minutes. 'Bit of a gruff old fellow,' says Craig. Having time to kill, he suggested that his friends take him to the Hilltop pub in

Omeo. If he had to wait to go to the doctor, he might as well have a couple of beers in the meantime as painkillers. His mates obliged, bringing out the beers to where he was lying in the back of the ute.

When he got to the surgery, the doctor said he wanted to cut the boot off his damaged foot. 'I said, "No, no, you're not cutting me brand new boot off."' So Craig got the nurse and the doctor to pull it off. 'It was like taking an anvil off the end of me leg, with that dead weight.'

His leg was put in full plaster right up to his groin. The full plaster made riding tricky. His mates had to help him up on his horse because he couldn't bend that knee. He'd let the stirrup right out and hung on to it with his toes to stop the stiff leg swinging about. But Craig Orchard still caught a brumby. Reckons he must be the only bloke in full plaster who's done it. It took him six months, though, to really get back into brumby running after that, to trust his leg, still weak after four months in plaster. Needed to get his confidence back in the leg and not worry about smacking it into a tree. But that was years ago.

As for Bogong Jack, he's still around. Craig handled him early on and thought of breaking him in but decided to keep him as a sire. He now presides over a herd of broodmares in the paddocks at Craig and Tahnee's farm. He's a real character, Craig says. Comes down through the paddocks to the house every now and then to see what's going on. Clears all the fences on the way down, gives the riding horses a bit of a hiding, hangs around for a bucket of oats then jumps all the fences and is off again. Sometimes he roams further afield, jumping the fences in the other direction and disappearing into the bush for a time. Craig will notice he's gone and wonder. But he always comes back.

2

A PONY CALLED PARROT

Ginger Soames was as excited as a seven-year-old could be, tossing and turning in her bunk, thinking about her birthday present: the black pony that was waiting for her down the road. It was four o'clock in the morning and she was in the Benambra Motel room with the rest of the family, who'd all arrived the night before for the early morning pick-up. Her five-year-old brother, Archer, had been pretty excited, too; he'd heard that Benambra was a town full of horsemen—'cowboys'—and was looking forward to seeing one riding a bucking bull down the main street.

The children's mother, Lisa, had felt mildly apprehensive, wondering about the brumby her father, Terry, had arranged for them to collect. Parrot had been caught years ago in the bush and ridden by two sisters but hadn't been worked for some time and Ginger had never ridden before. The family had driven down from Sydney to collect the pony sight unseen and eight hours was a long way to drive back if it didn't work out. But she trusted her father. Terry, a horseman from Tasmania, had broken in horses, some of them brumbies, and knew what made a good kids' pony. He knew Parrot's owner and if Jock Sievers said the pony was quiet, then that was good enough for him. Terry had been brumby running with Jock, helping him catch horses for his Australian Alpine Brumby Park and for rehoming. This one was a family favourite, a seventeen-year-old 13.2 hands high pony his daughters had ridden for years but outgrown. 'It'll be a good horse,' was all Terry said.

Ginger's wait was over at six o'clock. The family arrived at Jock's house and were taken by him up a track to a paddock where several horses grazed. Ginger was impressed with the man who gave them a demonstration of whip-cracking—he was 'one hundred per cent country', she says. But the little girl was more interested in meeting her future pony. Jock called out to the group of brumbies and Parrot came trotting over, a jet-black pony with a fluffy winter coat and hairy beard. Parrot looked like a yak, Lisa recalls, covered in the long coat he'd grown in the cold of the high plains. 'Is that him?' asked Ginger, looking excitedly up at her mother. If Lisa had any doubts about the pony, it was too late; Ginger had just fallen in love with him. He was the best horse she'd ever seen.

Although Ginger had watched her mother ride her horse at their home on the outskirts of Sydney, she wasn't sure what to do herself when she was helped into the saddle. She gave Parrot a little kick and steered, just as Terry and Jock had told her to do. The pony didn't take off or try anything on, he just did what he was asked. They loaded him into the horse float and took him home.

Ginger was ecstatic. She was so impressed with her new mount that she wouldn't put him in the paddock with the other horses when they got home but sat in the backyard watching as he grazed on the lawn. 'Mum, did you see that!' she'd exclaim, as Parrot moved a few steps, or 'Look at his eyes!' That was four years ago. She's still besotted.

Ginger learnt to ride around her home; in the backyard and on the few acres the family owned. Parrot was too strong for her at first and the slight girl found it hard to get his head up, Lisa recalls. 'This tiny little kid! All she wanted to do was go faster and all Parrot wanted to do was eat,' she says. Parrot would run along with his head on the ground trying to snatch mouthfuls of grass as Ginger booted him and pulled on the reins.

Within two months they were trotting and cantering to the letterbox, 'Ginger's little backside bumping out of the saddle'. Lisa would watch as her daughter came perilously close to falling off and as Parrot would slow down for her. Her first tumble occurred on the front lawn as Parrot stopped to put his head down and Ginger took a forward roll off him and landed on her bottom. It didn't deter her a dot.

Ginger and Parrot joined Glenorie Pony Club in Sydney near her home, picking up tips such as how to ride safely as well as riding skills needed for mounted games, jumping, cross-country and hacking. Within a year she was competing in every event possible including her favourite games: barrel racing and bending (cantering between upright poles without knocking them over), riding with more 'yahoo' than skill, her mother says. Parrot is speedy, for sure, Ginger says. At the time, Parrot, who'd had years of experience, was teaching her.

By the end of the pony club year Ginger's skill had caught up with her ambition and she scored her first ribbon at the club's district jamboree. The end-of-year jamboree is like a 'grand final day' for pony club, Ginger explains. After that she was taking home ribbons for mounted games, jumping, cross-country and even a placing in a club dressage competition. Ginger is officially in D grade for jumping but Parrot showed one Sunday morning at pony club that he could jump to B grade level. Pony clubs grade horse and rider combinations from E (beginner) to A in showjumping. (In New South Wales a D grade jump is 60 cm high.) The pair were doing the 'six bar' jump—jumping a succession of poles that get raised each round. As bigger horses clipped the poles or refused, Parrot cleared 1.5 metres.

Then Ginger caught up with Parrot and started to teach him. She taught him how to 'collect', although he sometimes resists, then how to perform a side-pass, leg-yield, roll-backs and hindquarters turns, training as much as possible. She goes to as many of the Steve Brady horse training clinics as she can to learn more about horsemanship. Parrot and Ginger found themselves enjoying the sport of campdrafting at her parents' farm at Bathurst; their next challenge is to compete in it. Parrot quickly showed that he had a feel for working with cattle and tries to round up the family's chickens, too.

Ginger says 'Pary', as she calls him, always looks after his rider. He'll charge around a barrel flat-out for her but won't even break into a trot when she puts her five-year-old brother Oscar on him. If she falls off, he comes over to her and nudges her until she gets up again. He knows when she's sad and if she's angry, she says, but he is always happy and relaxed himself. You can dress him up, get him to carry things, crack a whip off his back and he'll stay calm. When the local paper wanted a horse to appear in a photo walking next to a man playing bagpipes, Parrot obliged. It was Walk to School Day and although the local kids took other pets on the walk they didn't usually take horses because they took fright at the sound of the bagpipes. But Parrot didn't even lift his head from the grass he was eating when the pipes sounded and then walked along with his head low and relaxed, despite the deafening sound, carrying schoolbags the children had thrown on his back.

He's a character alright. The family used to have an orphaned lamb that Parrot would wander around with. Lambie would come if you called him, frolicking and kicking up his heels with Parrot doing little bucks around him. But he's definitely a 'real pig' when it comes to food, Ginger says. Potato crisps, banana skins, those 'sour' lollies. Shake a plastic bag in front of him and Parrot's nose is in it in a flash. He has worked out a way of getting under an electric fence for the grass on the other side, too. He sniffs it first, checking to see whether the hot wire is on or not, then twists his neck below it, gets on his knees and dashes underneath as quickly as possible because he knows he's going to get zapped.

Ginger once fed him some hay near the dam in his paddock but he accidentally pushed it into the water and they found him standing in the middle of the dam trying to catch strands of floating hay. Much to her mother's horror Ginger then started to play fetch with Parrot by throwing his food in the dam. Parrot loves a swim—it's hard to get him out of the water once he's in.

Because he's a brumby he's handier that most horses on trail rides. The country at the family farm in Bathurst can be rough: rocky, bushy and hilly with crevasses and ditches but Parrot takes it all on. If there's ever a difficult piece of track, Ginger points Parrot towards it first for the other horses to follow.

A brumby's heart is bigger than other horses, says Lisa, it will give you all that it has. Lisa adores her own two horses but says there'll never be another like Parrot. Ginger agrees.

Late in 2012 Ginger and Parrot competed at the end-of-year jamboree for her pony club division, up against riders from seven other clubs. They took home the Reserve Champion sash for horse and rider.

Not bad for a brumby from the bush.

Aleshia Lancaster recalls her best mate, Parrot

I broke Parrot in when I was twelve after Dad caught him at Native Dog Creek. He was so quiet I could crack two whips off his back at the same time. I used to get on him by jumping over his bum, and to get off him I'd just slide down backwards off his bum! Not long after I broke Parrot in, Tahnee and I went camping for a week out to the Limestone Paddock. She took Sindy the Welsh Mountain pony and I took Parrot. We night-lined them [a line tied from tree to tree with a rope in the middle] but the next morning they were gone! Craig was going to the rodeo at Jindabyne and found Sindy the white pony on the flat but no Parrot. Tahnee and I stayed there for a few more days but Parrot never turned up—he was off with the other brumbies. Oh no!

I remember one of the old cattlemen that lived on the other side of Limestone was a bit notorious for shooting brumbies and I rang him up and said, 'Please, if you see any black brumbies out on the edge of the road, don't shoot them until I catch mine.' I think he thought it was quite funny at the time.

A week later, after we had gone out every day looking for Parrot, Dad was sitting in the ute shaking a bucket of oats out of the window and—wow—Parrot stuck his head in the window. He had been flogged by the other brumbies but he was okay. What a clever little horse!

Safe and sound at home Parrot did cattle work and novelties and Tahnee did some brumby running on him. I remember he used to get into the oats in the shed and once ate the whole barrel; he lay on the ground and was so sick that we gave him apple cider vinegar to get the gas out of his belly.

After a few years I grew out of Parrot. Ginger, a little girl from Sydney, has taken care of my special brumby, and is going awesome on him. He will always be my best mate and the most enjoyable horse I've ever had.

3

THE MAN WITH THE HORSE IN HIS CAR

As a boy growing up in country New South Wales, John Stubbs would watch as carriage-loads of brumbies were taken off trains at the local rail siding and listen to the clatter of hooves as they were loaded into cattle trucks bound for the knackery. It was the early 60s and his family was living at Mulgrave, a couple of miles out of Windsor in the Hawkesbury Valley on the northwest outskirts of Sydney. The brumbies had been caught in Queensland and were headed for an abattoir nearby that supplied a pet food company. John was about ten years old and felt sorry for the horses, which were often left in the yards for days at a time with no food, water or shelter.

He'd always wanted a horse of his own but the closest he got was riding friends' horses before he moved to suburban Sydney, then paying to go on trail rides, as city people do. He took up big game hunting as a young adult, travelling throughout Australia to hunt and later going to Africa, New Zealand and Canada. Combining riding and hunting on a sixteen-day trek in the mountains of the Yukon in Canada in 1987 was 'like heaven'. Later in life John left Sydney to move to Omeo in Victoria's high country, hunting deer in the mountains surrounding the town in his time off and occasionally coming across herds of brumbies. He was out one weekend in 2001 looking for new places to hunt, when he came back with a little more than he'd bargained for.

He'd stopped at a likely location—a snow-grass plain at the edge of the state forest near Tumbarumba—and scanned the clearing through his binoculars for deer. As he let the binoculars drop something caught his eye to his right in the trees. He looked closer at the animal moving among the foliage. It was a foal. The small chestnut was walking around a snow gum suckling at some moss on the branches.

John scanned his binoculars in every direction looking for its mother. Not a horse in sight. He had seen a mob of brumbies a few kilometres away but this foal was on its own. He edged slowly towards it, keeping a big tree between him and it—the way hunters do when they're approaching game. The foal pricked up its ears when it saw him and looked cautious. John sat down on a log about 20 metres away from the gangly legged brumby, being careful not to look at the little horse directly and frighten it off. He made a nickering noise to it. The foal approached, step by step, and extended its neck and velvety nose to sniff the top of John's head. John slowly moved his hand up towards him. The foal startled and ran back to the snow gum.

John noticed that the foal was a colt, and the last part of its umbilical cord still dangled beneath it. Couldn't be more than a few days old, he figured. He remembered some milk he had in an esky in the old Range Rover and went back and got it, returning to where the foal was standing. The foal edged up to him again. Tentative. John dipped his finger in the milk container and held it out. The foal sniffed his fingers and started suckling. John stood up but as soon as he did the foal shot off. He sat down and offered the milk again; the youngster resumed its suckling.

'What do I do now?' he asked himself. Here was a foal, perhaps two days old, no mother, and no herd, stranded and in danger of being savaged by wild dogs or starving to death. John got up and headed for the Range Rover, thinking about his options. The foal followed, sucking on his finger. 'Looks like I'm taking it home.'

The rear seat of the four-wheel-drive was already down to allow for the times he slept in the back and the foal would fit in fine. He stood at the side of the four-wheel-drive calmly as John drew out a rope from inside and slipped it round his neck. Holding him by the rope John squirted more milk into the foal's mouth from a container he'd found in a first-aid kit. After twenty minutes or so, during which the foal had settled down after a small panic, John opened the back of the Range Rover, grasped him by the legs and lifted him in, putting him in the hollow behind the front seat. As he drove, he felt a small head nudging him under his left arm. Trying to get a drink, he says, laughing, thinking back.

As John headed home to Omeo, he stopped at a service station to fill up and was paying for the petrol inside when a woman came in and looked at him. 'You've got a horse in the back of your car!' she said.

John got the foal home and bottle-fed him, hand-rearing him and buying him a mare for company. Paddy grew up to be a stocky little bloke, 14 hands high, and smart. Much smarter than a thoroughbred, John says.

John started training him. Horse and rider had a lot of learning to do together. He asked around and read up, trying to understand what went on in the minds of horses. Better to understand them than belt them into doing something, he says. 'I've never hit a horse. Though I've yelled at them! There's no point getting cranky with them and belting them when they don't understand what you want them to do. Doesn't work and it's wrong,' he says. 'You've just gotta be patient till they trust you.' John does a bit of barefoot (no shoes) trimming for clients in the area—shaping the hooves to mimic wild horses' hooves and improve their soundness—and says you can always pick the horses that have been hit around the head or treated badly. You can see the owner's behaviour in the way the horse reacts to you.

Paddy turned into a great little bush pony. He has an incredible memory and sense of direction, says John. You can be riding him back to camp and turn him onto the wrong trail and he'll pause and look back, letting you know.

John now has four horses. The companion mare, Harley, was already seventeen when he bought her and died at 25, leaving a son, Scooter. John also bought a thoroughbred to ride. Then there's Mitta, the brumby mare.

Mitta had been caught as a foal on the Nunniong Plains, 30 kilometres away, and belonged to a woman who couldn't do anything with her. She was a good-natured horse but everyone in the area had written her off as a hopeless case, says John. He bought her cheaply, intending to train her as a packhorse, but she had a couple of bad 'wrecks'—broke the breakaway string when she was tied up once and took off, the packs slipping under her belly before she went through a fence in panic. For more than a month after that you couldn't even get a saddle pad on her. A few people said she would probably never be any good after the wrecks. It took a long time but bit by bit John settled her down.

He later went back to the woman he'd bought Mitta from, with photos of the horse standing quietly up on the Bogongs with the pack saddle on her back. The woman cried. 'You wait till I show them these photos—they all said she was no good,' she said. Now Mitta's the best little packhorse, and he rides her, too. She even got complimented by an old mountain cattleman John used to ride with.

John heads off on weekends, and whenever he can, riding Paddy, with Mitta in tow carrying his swag and supplies in packs. He goes all over the Bogong High Plains, up the divide between the Cobungra and Bundara rivers, Nunniong Plains, camping out, hobbling the horses in a taped area at night so they can graze. The horses know the drill.

Back at home they're cheekier. During the 2003 bushfires and drought, when there was virtually no grass around and no hay, John would let the horses into his garden to pick at whatever they could. Paddy learnt quickly how to mount the wooden verandah and flick open the rear sliding door. John recalls watching television one day, hearing a noise, and looking around to see Paddy standing behind him in the lounge room. As he was backing him out he noticed Paddy looking at the hunting trophies—the mounted heads—hanging on the wall. Another time Paddy snuck in to help himself to a bucket of apples near the back door when John was in the shower. The stains of the apple juice are still on the carpet.

The garden flourished after the drought and John kept a close eye on the horses whenever he let them into his backyard. He sowed 30 sweetcorn plants and watched on many occasions as the horses sniffed the plants and moved on, disinterested. The plants grew tall and sprouted corn ears. Still no interest. One night he left the horses in the backyard. The next morning, just after he'd woken up, he heard a crunching noise outside and went out to see Paddy finishing the last ear of corn. The horses had waited until the cobs were just right and eaten the lot.

It's not the only thing this cheeky chestnut's done but John Stubbs can only ever look back and smile, remembering the day he went bush and came back with a horse in his car.

4

SAVING THE WEST'S WILD HORSES

The plight of Australia's wild horses has touched the hearts and minds of people across the country, including those working to improve the horses' welfare under the umbrella of the Australian Brumby Alliance. The Outback Heritage Horse Association of Western Australian Inc. (OHHAWA) is one of the groups working under its banner. It was formed in 2005, after the rescue of some heritage horses living wild on Earaheedy Station near Wiluna in the state's mid west. A 'heritage' horse is defined as one that has originated from outback or isolated country, where horses have served people in different roles and where no breeding has been introduced since the early 1940s—something that can be determined from DNA testing or reliable anecdotal or historical, written evidence.

The association's members research and identify heritage horses and then, if necessary, remove and relocate them from remote areas in Western Australia, under veterinary supervision. They allow the horses to rest and put on weight before training them and selling them to good homes to help cover costs. The charity works with station owners, government departments and the RSPCA, provides advice to other horse rescue societies and associations, and lobbies to change the way wild horses are treated regarding capture and culling techniques. It uses passive trapping methods where possible, with the occasional ground muster to a safe trap area if necessary.

Here are some of the stories charting the progress of wild horses rescued by the association, adapted from accounts mainly written by its committee members.

Gunnadorrah Desert Jewel

December 2007

The first lot of rescued Gunnadorrah Station pintos have basically all settled into their new lives. 'Desert Jewel' or 'DJ' was one of the special cases. Some horses, like some people, do not deal well with trauma or stress in their lives. DJ was so traumatised by what she had experienced in the yards at the station that she was extremely nervous and distrustful of humans. Loaded into a truck with four other frightened wild horses, DJ was so weak she collapsed in the truck and was trampled by the others, resulting in raw patches of flesh on her wither and upper back. She arrived at Nannup, the location of one of the OHHAWA's rehabilitation centres, and was given a fortnight or so in the yards to settle down and get some basic feed and fresh water into her. Thought to be a young lead mare (DJ was around five years old), she was also a dedicated herd guard—a 'warning horse' who snorted like a freight train when distressed, instinctively alerting the other horses in her area.

Six weeks after her arrival, two of the other pintos had already been taught to lead in a halter and one had gone to his new home in Margaret River. DJ and another filly—Polyfilla—had been given the run of a long grassed raceway to put some weight on. DJ had calmed down slightly, but despite every precaution and gentle handling, she remained unable to cope with even minor pressure from human presence. On several occasions she became so distressed she almost caused herself serious injuries. All DJ wanted to do was get away—and it didn't matter if she had to throw herself over or through fences or yards to do it.

Even before her arrival in Nannup DJ had been purchased by a horseman from the eastern states who wanted her as a riding horse and, later, broodmare. It was realised, however, that DJ would have to be transported somewhere with safer facilities, like a padded round yard, before being properly handled. So Sheila, the association's vet, trucked her to Fred Watkins near Margaret River, an experienced horse handler who, along with Neil Innes, was helping Sheila train the wild heritage horses. Fred had already managed to calm another couple of older, traumatised Gunnadorrah pintos and had learned a great deal in doing so. Nevertheless, DJ gave him a couple of nerve-racking days when the serious job of getting a halter on her began. But during those initial intense hours, something amazing happened—DJ decided that she would finally try trusting a human. And she chose Fred.

Over the next few days DJ bonded with Fred in a way that the both of them had never experienced. She started to follow him around—even when he wasn't in the yards with her. She watched him when he was in the house with his family, her eyes following his every move. She seemed surprised that Fred was a decent guy who could actually be trusted not to frighten or hurt her. On top of this, he was a good leader, and all horses need a strong leader to follow.

To cut a long story short, DJ stayed with Fred. The horse trainer was so taken by this lovely, gentle-eyed mare that he spoke to Sheila, who spoke to DJ's owner, who very kindly sold her to Fred. He is going to keep DJ and ride her. It's also a real testament to the forgiving nature of horses—particularly heritage horses—and to the patient, kind training methods employed by Fred, that DJ has come so far.

2012 update

DJ is still with Fred and his wife Rachael, now happily settled into their new property near Perth. Over the years DJ has swung between tameness and her wild instinct. She wasn't put under saddle as the stress wasn't worth it. Possibly due to early malnutrition and the stress she was under for the first few months following her rescue, DJ unfortunately aborted her first foal and has also been unable to carry a foal to term, although the Watkins hope one day, with help, she may. She spends her time between Fred and Rachel's home and a nearby agistment property, where she can enjoy running with other horses in a larger area. She is content and still very devoted to Fred.

The Importance of Being Puzzle

Early 2008

Whether you believe in a higher power or not, sometimes the universe just seems to bring certain souls together for a reason. The second Gunnadorrah Station rescue included a pair of young colts—probably half-brothers and later named Puzzle and Chet. Both colts were very nervous and new to the world of humans. Although only just weaned, the colts' initial experience of people had not been good.

Chet went to a loving home with Sophie, an OHHAWA member, and for many months received a great deal of care and attention, going from nervous colt to calm and healthy yearling. Chet had an umbilical hernia which needed to be removed by a vet—a fairly common procedure—but he suffered an allergic reaction to the anaesthetic and died. Sophie's shock and grief was shared by the OHHAWA.

As for Puzzle, he went to a lady named Andrea, who loved her new colt . . . but someone even more significant was about to enter Puzzle's life, Andrea's friend Louise.

Louise, who was kind enough to share her story, suffered from recurring depression. In the past year, Louise and her family went through heart-wrenching tragedy when Louise lost her husband in a motorcycle accident. Only a few weeks afterwards the family pony, belonging to Louise's young daughter, died. A month later the property Louise was renting was purchased by developers and her lease was terminated.

Louise fell into a severe depression. She had trouble getting out of bed and her family and friends feared the worst. Then Andrea had an idea—she told Louise about Puzzle. She said that the newly acquired young colt, also rather lost and alone in the world, needed someone to spend time with him and tame him. Andrea claimed she was too busy and asked Louise to help. Louise agreed. It was an effort but almost every day Louise got out of bed, and went to see Puzzle. She sat in the yard and talked to the wild colt, sometimes for hours.

By the end of the first week she was sitting at his feet while he ate. And as Puzzle got calmer and tamer and healthier, Louise got brighter, too. Over the coming weeks Louise helped to train Puzzle, teaching him to lead, be brushed and have his feet picked up. She even taught him to 'shake hands'. Puzzle bonded strongly with Louise and began to follow her everywhere.

Andrea saw what was happening and made an incredibly generous and caring decision. She gave Puzzle to Louise. The unutterable joy on Louise's face was thanks enough for Andrea. Now Puzzle, happy with his new owner, is going from strength to strength. Just like Louise . . .

November 2012 update

These days Louise is a qualified veterinary nurse. She still owns Puzzle, now a six-year-old gelding. Louise says, 'He is beautiful, much loved, happy and healthy. Puzzle is very talented, going nicely under saddle and full of tricks.' He performs the Spanish walk, curtsies and lays down on command, and also loves to jump. Louise still regards him as her best friend.

(The names of the people in the story above have been changed.)

Darrah Dark One and Roscoe the Reverser

Darrah Dark One, they called her. Sheila the vet named her after a respected Indigenous woman in the area. Darrah was the odd one out, the only non-grey horse ever rescued from Earaheedy Station. And Roscoe the Reverser, the giant colt with the face only a mother could love, who got his name after he made quite an impression on Ross Quartermaine, the former owner of Earaheedy who helped Sheila with the rescue of the station's horses.

It was December 2005, after years of drought, and the station's wild horses were starving so the OHHAWA embarked on a second rescue. Two-year-old Darrah was captured via water trapping along with two orphaned fillies, a grey mare and Roscoe, a very big, very skinny colt about two-and-a-half-years old. Several others that Sheila and Ross managed to catch died, or were humanely destroyed, due to their poor health from chronic water and food deprivation.

Somehow, tough little Darrah and big Roscoe made it, probably due to their youth and incredibly hardy bloodlines. In Darrah's case, it is thought that one of the dominant grey stallions took a trip to the large, neighbouring station of Wongowol—the home of a big number of smaller, darker heritage horses also provided as remounts during World War I—and stole her (it is said that wild stallions only steal the best mares!). Or perhaps somehow Darrah wandered into Earaheedy on her own. We will never know.

It was obvious, though, that Darrah had been pushed around by other horses: she was very thin and carrying extensive bite injuries on her back when she was captured. Sheila could see that Darrah was a well-conformed, pretty horse, tough but quietly accepting and very intelligent. She was a survivor, and Sheila was optimistic the filly would pull through.

Roscoe was captured with a lovely steel-grey colt, a mate about his own age. Sadly, the other colt was in very poor condition, and collapsed not long after being trapped. Sheila and Ross spent hours in the dust and heat trying to save him. At one point, Sheila was on her knees beside the colt, while Ross stood above her, holding the drip that she was putting into the colt. Roscoe, on the other side of the pen, began to make his way over to them. Sheila and Ross were in an awkward corner, trapped between a water barrel and the stricken colt they were treating and now Roscoe, who suddenly stopped a few metres from them, turned around and began reversing towards Ross.

Ross, slightly nervous, made a comment to Sheila about the big wild horse backing up towards them; Sheila, intent on her patient, told him not to worry. But Roscoe continued to reverse, moving closer and closer to Ross, who was becoming very concerned. The colt was typical Earaheedy, about 16 hands, big boned and very powerful with huge hooves. Ross again spoke quietly to Sheila, who now glanced up and saw that the colt had reversed right up to Ross. 'Just stand still and hold the drip,' she said. 'It'll be alright.'

Then she got the giggles. Big wild horse reversing calmly up to big nervous man. It must have been quite a sight. Roscoe backed right into Ross, shoved his rump against him and started using Ross as a human butt-scratching post. It took a huge effort from Ross not to fall over (because Roscoe wasn't light), drop the drip or laugh, while Sheila kept giggling. When Roscoe was done with his scratching he moved away casually, having provided the former station owner and vet with a lighter moment in an otherwise sad day.

Unfortunately the darker grey colt died but Roscoe survived, and acquired his name, due to his 'bonding' with Ross and also just because Roscoe the Reverser sounded good.

That rescue was far from plain sailing. At one point, the station flooded—the deluge after the drought—and suddenly there was so much rain the badly deteriorated, dirt roads became impassable quagmires. The two fillies had been moved to a neighbouring station but the grey mare, Roscoe and Darrah were stuck out on the Earaheedy flats in the makeshift pens. Every day for three days Sheila and Ross drove many hours along a back route to get hay and fresh water to the horses. Finally the roads dried out enough for them to get in (again by a back way) with a truck. They constructed a makeshift loading ramp and, almost a day later, had the horses in the relative safety of the station. It was a big job.

After several weeks recovering in the yards, the Earaheedy horses had improved enough to make the journey south in March 2006.

Darrah recovered quickly, mainly due to her smaller size and youthfulness. She put weight on and her injuries healed rapidly, although a nasty deep bite on her wither took slightly longer and has left white scar hairs. The beautiful, intelligent young mare soon had a healthy shine to her black coat and was snapped up by a lady from Margaret River, who planned to ride Darrah when she was old enough. Darrah bonded with humans remarkably well and settled immediately into her new home.

Meanwhile, Roscoe the Reverser continued to surprise. After his initial casualness with humans, he took a little longer to settle than Darrah. His trust took longer to earn but Neil Innes, the natural horsemanship trainer from Margaret River, took his time with the big colt. When Roscoe was comfortable with the basics—the halter, leading, floating, having his feet picked up—he went to Nannup to join some of the other Earaheedy horses.

In Nannup Roscoe was further tamed and he blossomed. He loved affection, enjoyed his lessons and his food! And he grew, and grew, and is still growing. Like the other Earaheedy greys, with their big-boned Percheron ancestry, Roscoe has huge and powerful movement, coupled with a calm and intelligent attitude, ideal for horse sports like jumping. After being advertised for only a short time, Roscoe found a home: a mother and daughter from Geraldton drove down to view him, having read about this remarkable line of horses. They bought Roscoe immediately.

2012 update

Roscoe the Reverser and Darrah Dark One are both still doing very well. Darrah became a wonderful riding horse for her owner; however, after several years she developed back problems, possibly due to her old injury. The otherwise perfectly behaved mare started to pigroot regularly, unseating her owner on several occasions. Then her owner had to move interstate for work and Darrah was purchased by Kevin and Katherine in Nannup as a broodmare. She was five years of age, a beautiful horse in looks and nature; and now Darrah's a mum, too. Her first foal was a filly named Wanda, whose sire, Gunnadorrah Tom Thumb, a dark taffy-coloured stallion, was rescued from the pet meat station on the Nullarbor. Wanda is a *huge* personality, possibly because Darrah is a great, if somewhat 'hands off' mum. Her second foal, Firefly, was also a filly by Tom Thumb, and the same rare colour as her sire. Sadly, Tom Thumb died in 2011. Firefly went to live with Tom's owners, who adore her. These days Darrah spends her time being ridden (her owners use a 'treeless' saddle, which doesn't put pressure on her back) and living a quiet life in the paddocks of Nannup with other rescued horses purchased from the OHHAWA.

Roscoe has never had it so good, either. He was broken in, learning quickly and eagerly, then sold by his Geraldton owners when he was rising four to a family in Merredin who had *long* been wanting an Earaheedy horse. Megan and her mum, Robyn, absolutely adore the Earaheedy greys and later purchased Earaheedy Norma as well. Both horses have started competing in endurance riding and Roscoe, particularly, is doing very well. He has a naturally low heart rate and is a real 'cruiser', Megan says, looking after his rider and only occasionally trying to pick his own path and hop over fences! Because Roscoe and Norma are so large and so 'different' to the usual sleek Arabian-bred endurance horse, they often attract attention wherever they go and have become quite the celebrities.

Roscoe has decided that he really does enjoy domesticity and these days the huge, gentle gelding also does a bit of babysitting of foals and helps in the training of younger horses as well as participating regularly in endurance rides and workshops.

But every year when Roscoe and Norma visit Nannup for holidays, Darrah still bosses him about, despite being so much smaller. Roscoe the Reverser just takes it in his stride, as good-natured as ever.

5

RUMPY AND THE LAST RUN

Terry Fenton was barely a teenager when he became part of Tasmanian horse history, riding in the state's last brumby muster more than 50 years ago.

At the time he joined other riders called in to help round up wild horses from the Rushy Lagoon station in the state's northeast. The property was going to be sold and the owner wanted a mob of 200 or so brumbies removed from it in the two weeks before the sale. Most of the mob were stock horses but some years before the station owner had let go an Arabian stallion, Mustafa, to improve the herds and he'd sired some dishy-looking offspring. There were also a fair few coloured horses among them.

Terry still remembers the excitement on the first day of the big ride as he headed out on his own brumby with his uncle, and the 'great runs' they had galloping through the bush after the wild horses in the days afterwards. It was pretty rough country, Terry recalls, boobialla and thick scrub then sand dunes to the east.

At the end of the two weeks an auction was held to sell the horses, attracting buyers from the mainland as well as from all over Tasmania. 'They were a pretty good type of horse,' Terry says. A few remaining horses were shot, leaving just one horse behind, a red-and-white mare. The story went that the bloke selling the property intended to come back to get her.

The mare lived in the sand dunes behind Seven Mile Beach. Terry and a few men who worked at Rushy Lagoon would go out at weekends trying to find and catch her. She evaded them for years. She knew the country thoroughly and would gallop safely through tracts pitted by wombat warrens as the horses pursuing her tripped and stumbled to their knees. She would race through thickets of boobialla scrub that slowed down Terry and his mates; they'd have to dismount and unsaddle their horses to get through the scrub to the next open flat. 'And she'd be gone by then.' They nearly had her several times, herding her close to the station's stockyards, but she'd just beat them.

Terry was riding his brumby pony, Rumpy, 'one of the toughest little horses you could ever throw a leg over'. Rumpy was notorious for bucking first thing in the morning. Terry was prepared for this and knew how to handle him, but others at the station weren't. He would saddle up Rumpy, lead him out from the stables to the grass in front of them and say to one of the jackaroos, 'That's your horse.' 'Then we'd all get on our horses and whatever jackaroo got on Rumpy, he'd get bucked off—every time!' But after that Rumpy would go all day for you, Terry adds.

One day the riders got the red-and-white mare up from the sand dunes and thought she couldn't escape. She was galloping along a road with a rock bank on its high side. Terry was riding Rumpy 'full bore' on the other side of the road, and was galloping through an area of burned bush when Rumpy slowed. Terry pushed him on but stopped and got off to check him when it was obvious the horse couldn't go on. He saw that Rumpy had been skewered by a ti-tree limb—3 feet of intestines were hanging out of his side. Terry poked them back in the horse's stomach and later stitched him up and gave him penicillin, and two weeks later was riding him again. 'He was a tough little cookie.'

They eventually caught the mare by letting loose some other horses, which she joined, and herding them all into the stockyard. Terry and another horseman took her back to Launceston and tried to break her in, but she was too old by then so they sold her to a chap in Launceston who wanted her as a broodmare. That, as far as Terry knows, was the last of the Tasmanian brumbies.

PART II

Droving days and outback ways

6

ROCKET, AN OUTBACK LEGEND

By Jeff Hill

Jeff Hill and his wife Cooee live at Six Mile Springs, a property just out of Maryborough in Queensland.

At the age of 76 I have worn many hats in the pastoral industry in the Northern Territory and Queensland: drover, head stockman, contract horse-breaker, station manager, station owner and cattle buyer. Over the years, I've owned and ridden many good horses, including terrific night horses and camp horses, but by far and above the best horse I ever owned was a little horse called Rocket.

My father bred ponies at his dairy farm at Biloela, 160 kilometres south of Rockhampton, Queensland. There were eight of us kids and everyone wanted a pony so he decided to breed them for us. He had a stallion called Young Ludo, one of a breed of ponies from Honolulu, a creamy horse with a great long forelock and great jumping ability. You could crawl under a fence and stand on the other side and click your fingers and Young Ludo would jump over it. Rocket was one of the ponies Dad bred later, Young Ludo's grandson, and stood at 13.3 hands, bay in colour with a star and snip.

In 1964, together with my wife, I was managing a property in the Northern Territory called Elizabeth Downs in the northwest of the state—3300 square miles of unfenced country. We always took our annual holidays during the wet season while no cattle work was taking place, so we went home to Biloela on holidays with our two children, Linda, aged four, and Geoffrey, two-and-a-half. They had their first ride on Rocket that year and my father gave Rocket to them.

The only trouble was that Rocket was nearly 5000 kilometres away by road.

The following year, a mate, Jim McGorry, and I flew to Brisbane to visit a chiropractor. Jim had a crook knee and I got migraines, though I found out later these were from drinking beer. I had my last beer eighteen years ago. Another mate, Cec Roberts, who was managing Tipperary Station, had a Volkswagen in Brisbane and wanted me to take it back to the Territory for him. I decided I'd also take Rocket back to the Territory for the kids to learn to ride on. The trip up towards the north of Queensland and through the Territory to the station, southwest of Darwin, was more than 4800 kilometres.

I loaded Rocket onto a rail wagon in Biloela, destination Cloncurry, northwest Queensland. McGorry and I unloaded him in Rockhampton and Townsville for a drink, a feed and a walk around. The road from Townsville to Cloncurry in those days was all black soil and this was in the middle of the wet season. Black soil is shocking when it rains—it clogs all the wheel wells and you can't drive through it. We made it as far as Richmond and the road was cut; however, the rail was open. While having a beer in the Mud Hut in Richmond we saw Rocket go past on the train. My wife Cooee, Cec Roberts and his wife were to meet the train in Cloncurry to unload Rocket. However, because of the wet season I didn't know if they had reached Cloncurry (long before mobile phones) so I caught a plane to Cloncurry, just in case they hadn't. Cooee and the Roberts family had made it through and had unloaded Rocket off the train. Jim McGorry arrived with the Volkswagen at four o'clock the next morning.

Getting Rocket from Cloncurry to Adelaide River—the town nearest Elizabeth Downs—was uneventful, but the big hurdle was yet to come.

Elizabeth Downs was on the southern side of the Daly River and the Daly was in full flood. I knew it was impossible to try and swim the Daly. At Adelaide River the meeting place is the bar and I went in there to see if I could find someone to help me get Rocket across the river. A chap who used to do the grader work for me, Fred Pocock, walked in. Fred had a 5-metre wooden punt so I thought, 'We'll put Rocket in the boat'. Everyone thought I was a bit crazy. The Daly's a big river and it was pretty rough. But I knew that if I could get him in the boat, he would stand up okay. If I was there with him, Rocket would do anything. We had no trouble getting him in the boat and Rocket just stood there, bobbing up and down in the boat as if he had done it a hundred times.

After we unloaded Rocket on the other side of the river, he still had 35 kilometres to go to Elizabeth Downs. I asked one of the men working for me, Rankin Liddy, to jump on him bareback and ride him home. Rankin looked at me and said, 'Yellow Creek, Christmas Creek and Saddlerail Creek will all be a swim for this little horse.' I told Rankin just to give him his head and he would be okay—I had to return the horse float to Adelaide River and then fly home.

When I arrived home Rocket was in the stable having a feed. Rankin said to me, 'That's the gamest little horse I have ever ridden.'

So Rocket's life in the Territory commenced. He was about five years of age at this time.

For the next couple of years the children learned to ride, how to put a bridle on Rocket and with much difficulty saddled him up—Linda was aged six, and Geoffrey four-and-a-half, so it was hard for them to reach the straps.

In July 1967, we left Elizabeth Downs and went contract mustering. A friend of ours had a farm at Adelaide River where no one was living so we used the buildings as a main base. My wife made some bread and a fruit cake and left them on the table to cool one day; on returning there was Rocket with the fruit cake in his mouth taking half of it in one bite—he had walked through the open doorway. My wife gave him the rest, but he wasn't too popular for awhile.

He would eat anything I would eat and loved ice-blocks, and is the only horse I have ever known to eat corned beef. He could almost talk.

Geoffrey and Rocket were inseparable, and when Geoffrey was about six the ringers took him and Rocket mustering with them. It was a ten-day camp and the camp truck came back into the station one day with a message from Geoffrey—he and Rocket wanted some cake. The ringers thought the world of young Geoffrey.

Rocket had a heart as big as a lion and would have a go at anything I asked him to do. He could jump his own height—4 feet 10 inches or around 150 centimetres—and I used to go exhibition jumping on him without a saddle or bridle. The exhibition rides came about at the Adelaide River Rodeo one day. I'd just been in a jump event when I said to someone, 'He'll do anything I ask him to,' and was dared to jump the course without a saddle or bridle, which we did. The jumps then were two bales of hay with rails on top of them. Until the children learnt to ride, I used to compete in open events on him myself and I must say he was never beaten in jumping, flag races, bending races and barrel racing. Rocket competed in venues from Timber Creek on the western side of the Territory to Borroloola on the eastern side. Darwin, Katherine, Adelaide River, Daly Waters, Mataranka and Larrimah were all places where he always met strong competition. We just used to stand him up on the back of a long-wheel base Toyota and travel hundreds of kilometres to these events.

I remember him one time in the sand at Mataranka, barrel racing. A chap named Jim Bohning owned a good all-round horse called Stockman. With about 30 horses in the event, Bohning and I dead-heated for first place on 17.2 seconds so we had to run it off. Bohning went out first with 17.2; I went out, 17.2. I knew the sand was going to beat Rocket—Stockman was a big horse, 16 hands, so could get through it better. So Bohning went out again with a smile on his face—17.2 seconds. While he was doing the course I pulled the saddle off Rocket, lightening his load, and rode him bareback—17 seconds and won the event. Bohning threw his hat on the ground and said, 'How the hell do you ever beat that little horse.'

By 1972 the kids were old enough to start competing. Geoffrey's first ride on Rocket in competition was at the Adelaide River Gymkhana. It was the under-10 flag race and he won a red ribbon. From then on all ribbons were blue.

The horse had a cast-iron will. On a sports day Linda and Geoffrey would ride him in their age groups and I would ride him in the open events and there was always a lady rider with her hand up to ride him in the ladies' events. He won hundreds of ribbons and trophies against top horses twice his size.

He became a pet to everyone at the shows and rodeos. He could undo just about any gate catch and get out and go from camp to camp, where he was given a piece of bread or cake—everybody gave him something. Then he would come back to our camp looking pleased with himself and have his main feed.

By 1976, Rocket was fifteen or sixteen years old and although he was still winning, I thought it was time to pension him off. My sister, who still lived in Biloela, had a daughter who wanted to learn to ride, so Dot and her husband came up to the Territory with a horse float and Rocket made the long journey home, where he was eventually put down aged 25.

He was a remarkable little horse and although it is over 40 years since Rocket made his mark in the Territory, people still talk about him—he became a legend. There will never be another horse like Rocket.

7

THE BUSINESS OF DROVING

By Ian McBean A.M.

Ian McBean owns Bonalbo Station in the Douglas Daly region of the Northern Territory with his wife Kay. In the past, Ian travelled vast distances across the top of Australia as a drover, at times moving stock through properties he was later to own. At the heart of the drover's livelihood and existence was the 'horse plant', and Ian explains how it worked.

There was a big difference in droving when I was involved in the 50s and 60s to the present day. As drovers our business was to walk cattle from one place to another—and that could be thousands of kilometres—and we relied on horses to do the job. These horses were always referred to as our 'horse plant'.

Today, drovers have a horse truck and probably a caravan or mobile home. They have four or five horses and quad bikes. There are fences on each side of the stock route and they put up electric-fenced yards or 'breaks', as drovers call them, to hold their stock at night.

In our days there were no fences or yards along the stock routes. The stations we passed through weren't fully fenced, either. Cattle could roam for hundreds of kilometres if they got away. As one of my old mentors would say when we were working in the western Northern Territory, 'Don't worry if they get away, salt water will stop them.' He meant the Indian Ocean, about 900 kilometres away!

My first trip, as a boss drover with my own plant, was in 1956, from Brunette Downs station in the Northern Territory to Springfield Station near Quilpie, southwest of Queensland, then out to Betoota, near Birdsville, and back to Quilpie with fat cattle to put them on the train to Cannon Hill saleyards in Brisbane. From there we went down to Courallie Station near the South Australian border then back to Quilpie again before heading back to the Territory. This trip took more than twelve months and our horses would have covered more than 5500 kilometres by road. Needless to say these horses would have travelled twice this distance while working the cattle, going in and out to different waters and feeding at night.

In the ten or so years of droving that followed I would go on many trips that lasted four or five months, after which I would walk the empty plant (no cattle) back to my home in Camooweal, in northwest Queensland. In the early 60s I walked cattle from Auvergne Station near Timber Creek in the Northern Territory into Queensland, a distance of about 2000 kilometres. It was written into our droving contracts that we could not travel more than 112 kilometres per week with cattle and we would do about 32 kilometres per day travelling with an empty plant. Whether you had cattle or an empty plant, you still had to keep all the horses shod, hence all boss drovers and most of the ringers were pretty good farriers.

The average size of a mob of cattle in the Northern Territory or western Queensland in that era was 1200 to 1500 head. To handle these cattle a drover would usually have 30 to 40 horses in his plant, even if he had a truck. These horses consisted of three different categories: the day horses that the ringers (stockmen) rode during the day; night horses for watching the cattle at night; and packhorses. Even if the drover had a truck he would still need packhorses because he was so far away from communication of any kind, be it phone or wireless. If your truck broke down or it got too wet to travel because of flood, you still had to keep the cattle moving, so you would transfer all your necessities—tucker, swags, shoeing gear, etc.—on to your packhorses in order to keep travelling.

The horses were looked after and cared for by the ringer known as the 'horse tailer', who would be in charge of them, usually for the whole trip, whether that was for a couple of weeks or three or four months. A ringer usually had three or four horses for his own use, and would ride a different one each day, rotating them, so every horse had at least a couple of days spell before being used again. In a droving camp a ringer could ride the same horse all day as they weren't working hard, but in a station stock camp they would change horses at lunchtime, especially if the cattle were wild, as these horses might be doing a lot of galloping, pushing cattle around and concentrating on holding the wilder ones in the 'coachers' (quiet cattle) to settle them down. Most cattle that were taken on long trips were reasonably quiet as they had been mustered on horseback, yarded up and worked in yards at the station before 'they went on the road' (the drovers' term for the start of a droving trip).

Droving was a 24 hour per day job as we had to 'watch' our cattle at night, using the 'night horses' to hold them on camp. The night horses were every boss drover's pride and joy. They were started off as day horses after they were broken in and after a few trips you could pick out the best of the day horses and start working them at night. They had to be really alert all the time and show an interest in what they were doing, pricking their ears and watching what was going on in the mob, such as cattle moving about or trying to 'feed out', that is, graze away from the group. A good night horse could be on one side of the mob on a pitch black night but would notice two or three beasts starting to feed out on the other side 200 metres away. The horse would immediately start to walk faster and get round the other side of the mob and turn them back in without much instruction from the rider.

When the ringers got the cattle to where they were to camp for the night it was reasonably easy to make them pull up and stay in one place. However, if the cattle weren't used to being held at a camp, the station manager would lend the drover a few ringers to help cover this settling-in time. This let the drover have two, three or even four ringers on watch at night until the cattle were broken in to being held on camp. Some managers would give the drover delivery two or three days before a full moon so by the time the extra ringers went back to the station the cattle had settled and one ringer could watch them. A good full moon was like daylight.

Where there were a few trees about, the cook would be instructed to camp near a tree where the night horses could be tied up. If there were no trees, like across the Barkly Tablelands in the Northern Territory, you would carry a star picket on your truck or if you had packs you had to carry a couple of long sticks you could lash together to make a stake. If there was a reasonable amount of timber, the horse tailer would be expected to make a windbreak with bushes for the night horses in really cold weather. He would also put a horse rug on them over the top of the saddle, with a surcingle round the rug to stop it flapping or falling off. The horse tailer would also be expected to make a fire during the day with plenty of smoke so the plant horses could stand with their heads in the smoke to keep the flies out of their eyes.

When the cattle were approaching the cook's camp in the late evening, the horse tailer would have brought the horses up close, be it packs or a truck, and started hobbling them so they couldn't wander off overnight. After doing this he would catch the night horses to be used that night. If there were three men on watch, he would catch six horses; three would be on watch and three tied up ready for the next watch. This meant the night horses had a good spell between nights of work.

The number of men in the camp would usually be six; the boss drover and three ringers with the cattle, a horse tailer and a cook. As the cattle settled down to being held in camp and were able to be watched by one man, the boss drover would usually do two 'dog watches', as they are known. This was when the cattle first go on camp at night and again in the morning while the ringers were having breakfast before daylight. The rest of the night would be evenly divided among everybody. Darkness would usually be from seven to five, hence each watch would be two hours with the horse tailer doing first watch and the cook doing second and probably the boss drover doing midnight watch.

The bullocks could spook at anything at night. Often you couldn't even tell what it was—a hawk or noises in the camp—you wouldn't know as it would happen so quickly. It was as if every beast hit its feet at the same time. The boss drover would have told all the men in the camp, including the cook, not to shake a blanket when they were rolling out their swag, not to rattle a billy can or camp oven and generally not to make any sudden noises that could frighten the cattle.

If something frightened the mob and they did 'rush' (stampede) then a really good night horse was essential. The sound was deafening and there would be a terrific lot of dust and it was very hard to see what was happening and where you were going. The riders would be concentrating on getting to the leading cattle and turning them around so they could 'ring them up'. This meant bringing the leaders round to run into the 'tail' (the last cattle in the mob), so the whole mob would be galloping in a big circle. At this stage the ringers would move away from the mob and let them slow up and settle down. Sometimes the cattle would only 'jump' camp and run for a few hundred metres, other times they might go a couple of kilometres before we rung them up. The ringers would be concentrating on watching the cattle and would rely absolutely on the night horses to jump over logs, holes, rocks or anything that was in their way, including dodging trees. If the horse fell, there was a chance that the cattle could gallop over the fallen rider or he could be severely injured by his fall. The rider's life depended on that horse, which is why a really good night horse was such a prized possession. As a drover relied so much on his horses he would keep good horses, especially night horses, for the whole of their working life.

The third component of the horse plant was the packhorses. Most of us could not afford a truck when we started out as drovers so only had packhorses to cart our gear and keep our business productive. These were sometimes mules and donkeys, which were very strong and were handy to drovers as they had very hard hooves and did not need to be shod.

The average drover on these long trips would have eight to ten packhorses to carry all their gear and swags. A pack consisted of a pack saddle with hooks on each side onto which the pack bags were hung. The pack saddle was held onto the horse by a front and flank girth and in mountainous conditions a breastplate and crupper would also be used. These packhorses would carry about 60 kilograms each. It was essential that the pack bags on each horse were very close to equivalent weights on each side, so the pack would not roll.

One pack would be for salt (to corn the beef as you had no way to keep fresh beef), another was for flour, another was your meat pack and there was a pack especially for horse-shoeing gear—shoes, nails, tools and a very small anvil. There were also a couple of packs for tinned vegetables, dried fruit and other dry rations. There'd be at least one canteen pack with a water canteen of 20 litres hanging on each side. Last but most important was the cook's pack where he had his plates and cutlery, some of his cooking gear, and all his cooked tucker.

On top of the bags, each pack would have one of the ringer's swags over the top and partially down each side, held down by a surcingle. On top of the swags we would put billy cans and Bedourie camp ovens, which were tied by side straps on the pack saddle. These ovens were made of flat steel, not cast iron, so they could not be broken. Most drovers would have an ex-army .303 rifle tied on with the side straps behind the swag for destroying injured stock or getting a 'killer' (it was written into every droving contract that the drover was entitled to kill a certain number of cattle for beef while on the trip). Everything had to be tightened in case you had a packhorse that took fright or bucked with his pack.

The first-aid gear we carried was minimal—a few bandages, a bottle of iodine, a few Aspros. Under present-day occupational health and safety we would have been breaking the law all the time!

So now you know how a drover's plant worked back in the 50s and 60s, here are some anecdotes from that time.

My first droving trip was as horse tailer for Max Shepley, taking cattle from Elkedra Station to Alice Springs in the Northern Territory, about 400 kilometres. As horse tailer I had to get the horses in at the start of the day, undoing frozen greenhide hobbles at four o'clock in the morning—the temperature drops below zero in Central Australia at night. This was a short trip of only a fortnight; however, it turned out to be one of the worst trips I've ever had. The cattle rushed every night and one night after we had brought them back to camp we realised Max wasn't with us. We went out looking for him and found him crumpled up by a bronco yard. The cattle had galloped past the yard, which was made of very thick wire and wooden posts, and in the dark and commotion his horse had run into the wire and Max had gone over her head and landed on a wooden post on his stomach and pelvis. How he didn't break his pelvis I don't know. His horse had hit the wire above her nostrils and ripped the skin off right up to her eyes. Max stitched her up and she healed but was badly scarred. Max couldn't walk and travelled in the truck (luckily this wasn't a packhorse camp) but came good after a few days.

On another trip, in 1956, we got to the Diamantina River in western Queensland and found it in flood so we swam the horses and bullocks across it. We were lucky there was a boat to take our packs over. Despite it being very wet we found a good camp on a sand hill but during the night the cattle were very touchy and we discovered we had landed in the middle of a rat plague. After a few nights, however, the bullocks got used to the rats and you could see a bullock stretching out his nose to sniff a rat while still lying down.

In those days there were stock inspectors continuously patrolling the stock routes, and sometimes if you had an outbreak of a disease like 'three day sickness' (bovine ephemeral fever) the stock inspector would get you to pull up at a quarantine reserve until your cattle were right to move again. These 'stockies', as they were known, were very helpful to all drovers as they were continually moving up and down the stock routes, and if anyone was in trouble they would do what they could to help you out. On one occasion I had some Aboriginal stockmen walk off camp during the night, taking their swags with them, which left three of us with 1500 bullocks and 40 horses. Luckily, a stock inspector by the name of Jack Travers caught up to us a couple of days later and was able to get us some more ringers.

The Victoria River at Timber Creek in the Northern Territory is a tidal saltwater river full of crocodiles, which can be crossed at low tide with cattle. I still maintain that a lot of those ringers would ride across the river with their feet nearly on the pommel of their saddle when moving cattle off Bradshaw Station for fear of the crocodiles.

When watering cattle in the big rivers in the Channel Country of western Queensland, you would have to let them drink in small mobs in case the leaders decided to swim over the river and take a mob with them. As some of these waterholes were up to 200 metres wide, it was a real hassle for the ringers to swim their horses over and bring the cattle back. Sometimes you could throw a stone out in front of the leader to make a splash in the water and turn them back, but you'd have to have rocks handy and someone who could throw a long way.

Droving for me ended in 1964 when I was lucky enough to draw a station in the Victoria River district in a land ballot; however, at that time long-trip droving was on the way out as roadtrains had come on the scene and were slowly taking over. These days I look back with nostalgia on that period of my life as I found it very rewarding without the hassle and stress of modern-day living and business. However, one must move with the times, which is what I have attempted to do.

8

NIGHT HORSES REVEALED

Allan Andrews spent his early years in Dirranbandi, Queensland. He worked with sheep as a jackaroo at Winton and Quilpie, with cattle for two years near Clermont, then went on to the Gulf around Normanton and Burketown in the late 60s and 70s working as a head stockman and later managing properties. He and his family shifted to the Northern Territory in 1981, managing one property for twelve years and the next for sixteen. Allan is now semi-retired and involved in environmental activities for the company he once managed, and lives near Darwin. He writes about his observations of horse behaviour, and about his own horse, Westaway.

After the period of hobbles and open paddock horse tailing, but prior to the days of hand feeding, trucks and horse floats, the working horse plant was run in smaller paddocks of, say, 4 square miles for convenient access.

The night horses were kept overnight in a small enough paddock to be supposedly caught in the dark on foot, but they were experts at camouflage and avoidance of the job ahead of them. They would stand very still in the deepest shadow, hiding behind any available bush, tree or structure, and always at the extreme ends of the paddock. Once saddled, though, they took to the chore with exuberance and almost always had mouths of steel.

The horse plant depended on the size of the camp and was usually in the vicinity of 30 to 80 horses. These mobs broke up into their own social standings within the paddock, for example, the elite open drafting horses were always in one group, the colts another and the normal plugs and mustering horses again apart. Another group was also separate: the rogues, 'rooters' (horses who always bucked) and the 'yangs', recalcitrant horses of questionable ability and usefulness. The yangs had a dislike for humans for whatever reason and were always waiting for an opportunity to hurt someone or throw them off. A number of good horses are guilty of the above but the good outweighs the bad. Often a horse will throw you just to prove a point or because it's being playful, is sensitive to the girth or full of energy.

The social order I've described among the horse plant was repeated when the horses were yarded, unless it was disrupted by man. One member of each group had a bell with a different tone strapped to their neck for identification at a distance in the dark.

The horse tailer would be out at four o'clock to ensure the plant was in the yard well before daylight so the ringers would have their horses caught at first light ready to go. It was not on to be late. A young ringer would be sworn at by the boss plus his workmates if they were late. So as a tailer you'd be out in the night-horse paddock in the dark searching for an elusive shadow, with no sounds from this shadow, falling over logs and the stumps of ant beds or down holes, sometimes in the freezing cold. Horses have individual gait, character, stance and habits so you use this knowledge to identify your horse in the dark because colour or markings are useless, and you know your horse for the day will be with his mates. You find one, you find the others. Success! You see the shape but on discovery it moves away, especially so if you lack experience. Finally, you manage to get the bridle on the horse, hoping the glow in the east does not appear, which means you will be late.

Saddled, the horse at least has no buck this morning. You move through the pitch black, the horse's ears pricked, trotting across the timbered creek trying to find the rest. You stop and listen. No sound, not even the slightest tinkle, so you trot a bit further into the void, trusting the navigation of your mount. Then one belled neck makes a mistake, a tinkle breaks the silence, followed by the rest of the bells, and the quiet of the morning is suddenly a bedlam of sound.

The night horse takes the bit while the whole plant bolts for the yard and you dare not be far behind—if you're not with them they will run back out. At a fairly respectable speed on a hard-pulling horse you try to remember where the overhanging branches are and to make out the faint line of trees along the creek.

Horses yarded, you shut the gate. That job done for the morning, you have a quick breakfast while the rest of the camp are saddling up then rejoin the musterers.

Experience and a calm approach are the keys to success when you're catching horses. Horses have a mistrust and are nervous of strangers. They are very sensitive to the emotions of humans. Horses can sum up how a person is feeling instantly and detect the changes in personality as they are exposed. They can detect fear, anger, inexperience and people they can manipulate. The 'nice horsey' attitude without authority will make life very uncomfortable for the rider.

Before attempting to approach a nervous horse your mind and posture has to be free of aggression and tension. You need to be relaxed, in other words, let the wind out of your guts. Nor should you engage a horse with direct eye contact; this works opposite to the message you are trying to convey. As a ringer you never raise your hand towards the horse or make clicking sounds. Anyone with quiet, stabled animals would have difficulty relating to the above, but this was the norm.

In hobble out-camps, especially with Aboriginal stockmen, horses were caught by the hobbles so it was necessary to get down to ground level on your haunches or knees and virtually crawl towards the horse's front feet, then quietly rise up the shoulder to attach the bridle. This done successfully, you then release the hobbles, which the horse originally thought was the plan.

Like most animals, horses are not social as far as humans are concerned, compared to dogs and cats. However, with close continual association they can become close. And this is how it was with Westaway.

Westaway

I purchased this chestnut from the Westaway family in the Queensland Gulf north of Normanton on the Flinders River in 1974 after he was demoted from a stallion to a gelding. After experiencing the productive life of an entire, though, it seemed he could not accept the demotion. There are stories of Westaway cantering around the front of a vehicle trying to prevent this machine taking his latest girlfriend away.

His intelligence and learning abilities were exceptional. It still pains me to realise I did not have the ability to understand him like he understood me. Possibly his analysis of me was not always complimentary.

At Kynuna Rodeo he escaped from his stall and successfully crawled into the stall of pony belonging to a prominent lady, completing his intention of helping the pony devour its feed. After being discovered by this lady, who became vocally distraught, he calmly crawled back out and disappeared into scrub on the banks of a creek next to the stalls.

He was 15 hands high and his ability was exceptional. As an all-rounder he was placed first at least once in every event in the rodeo arena. He won his last open draft at the age of 22 years. His prowess as a pick-up horse dragging broncs twice his size from the arena back to the unsaddling chute was widely acclaimed. His pick-up partner was a horse called Coal Black, owned by Ian Rush. Coal Black was, in my opinion, one of the best and safest horses for taking a rider off a bronc. Immediately after successfully retrieving the rider, Coal Black would stop and turn away from the buckjumper so the person who'd just come off him would not be kicked or trampled.

When Westaway retired he was fed and stabled every night except for one occasion when friends called in to stay the night with two of their own horses. For convenience we put one in Westie's stall and fed him outside. He was so offended he tried to attack the offender over the top rail with no success then completely 'threw the teddy out of the cot', promptly ignoring his feed then disappearing for two days and refusing to accept any condolences on his return.

This episode slammed home the extent of the damage that can be caused to a horse's feelings. He assumed it was his right from services rendered to claim his place in life.

He whinnied to me one day from a hundred metres away. Thinking this was unusual, I walked over and discovered he had a piece of steel penetrating the frog of his foot. He never moved while I prised it out yet he was sensitive around the feet while being shod.

People underestimate the communication abilities of animals with themselves and with us.

9

THE BREAKER, THE BRUTE AND THE QUEEN

It was late afternoon on Barkly Downs Station in the Northern Territory, sometime in summer 1960. Most of the station workers were gathered around the cattle yards, ready to watch a young Aboriginal stockman take on a rogue horse. The horse had been giving a few of the ringers grief and Lurick Sowden, in his teens but already a fearless horse breaker, said he would ride him.

'Everyone came out to watch,' recalls Maxine Holt, who was to marry Lurick. 'All the families who worked there, the manager, the overseer, the bore man, the butcher, the women who worked in the kitchen, the kids and the fellas were all down at the yard watching.'

The station workers staked out their positions looking through the rails of the yard or perching on the top bar of the fence ready for the action. Some of the women, including Maxine's mother, were a bit scared for Lurick. They should just let the horse go, her mother said. A couple of the older women told Lurick he was a silly bugger.

'He's a devil,' someone said of the horse.

Maxine remembers the overseer, Laurie, warning Lurick not to let the horse out but Lurick insisted. 'We're going to have to let him out 'cos he keeps going silly in the yard,' he said. The men had a lot of trouble saddling the horse and it was still mucking around when Lurick took his chance and jumped on and asked for the gate to be opened.

'Are you sure?' asked Laurie.

The horse saw the gap to freedom and shot out of the yard, bucking furiously, twisting and turning, then thundered off across the plain.

The crowd watched and shook their heads.

'He won't come back alive.'

'That horse will put its foot in a hole and that'll be the end of him.'

'He's a gonna.'

They all watched and waited, heads turned towards the west as Lurick and the horse merged as one, a lump on the landscape, then became a speck, then as they disappeared from view.

Everyone stood round waiting, not sure what to do. 'We might have to send someone out looking for him,' one of the women suggested. An hour or so later, they were still standing around talking and scanning the horizon. Nothing but Mitchell grass and scrub.

Then, 'Here he comes!' one of the ringers called out. 'He's coming from the other direction.'

Everyone turned round behind them as Lurick and the horse appeared, a huge semicircle away from where they were last seen, walking towards them. Horse and rider were plastered in sweat. 'That poor horse could hardly walk when it got back,' says Maxine.

'I took him for a long gallop to settle him down,' recounts Lurick. 'Galloped the living daylights out of him.'

Lurick let the horse run itself out then made it run some more.

'He turned out to be a real good horse,' he says.

The story is still talked about today.

In his day Lurick Sowden was known as a gun breaker and horseman. He was born the son of a white drover and an Indigenous woman from the Wakya people on Alexandria Downs Station in the Northern Territory or, as another version has it, on a droving trip. He learned to work horses and cattle at stock camps as a boy but left the station at sixteen, not liking the way the Indigenous people were treated there. He moved around between several stations in the Territory, Queensland and Western Australia, breaking in horses, mustering and droving, working at Barkly Downs as a contract breaker with Ian McBean (see Chapter 7) at times when he was about seventeen or eighteen. The two men broke in seven horses a week—each—usually by 'bucking them out' or staying on until the horse tired.

Maxine Holt first met 'Soda', as her cousin George Rankine named Lurick, in Camooweal in Queensland, then again when she was at Barkly Downs on holidays visiting her mother and stepfather, who were working on the station. Maxine had relatives from the Wakya people, too. She'd heard stories about Soda from her cousins on Rocklands Station, and knew he was daring, unafraid of any horse. She says that because he had come from the Territory it was hard for him at first to break into the circle of youths his age at Camooweal, some of them her relatives, and who'd all gone to school together. When they first saw Lurick he was riding a horse that was bucking. 'Whenever he knew he was about to be bucked off, he'd jump off the horse and land on both feet,' she says. 'And when he did that, they thought he was a show-off.'

But Lurick was a person who could get on with anyone, she says. They married in 1964 and lived in Camooweal, where they had three daughters affectionately known round 'the Weal' as the 'Soda Pops'.

Despite his skill in riding out a bucking horse, Lurick never rode broncos at rodeos—but he does have the distinction of being the first man to stay on the famous Mt Isa Rotary Rodeo 'feature' horse, the Barkly Brute. The late country-and-western singer Stan Coster, the 'Cunnamulla fella', wrote a song about 'the mighty Brute' in 2007 that begins:

In north western Queensland there's a wealthy copper town
And its yearly Rotary rodeo has brought it world renown.
And this song is dedicated to a cyclone from the chute,
A vicious chestnut gelding known as the Barkly Brute.

The Brute started out in the novice saddle bronc in 1967 after the more famous Spinifex was 'bucked out'. He became the rodeo's feature horse in 1968, then again in the next two years, and appeared in front of the Queen in a special rodeo staged for her in 1970.

But Lurick rode the Barkly Brute before he became a rodeo horse. In fact he broke him in at Barkly Downs. He doesn't remember him as being any worse than other horses he was breaking in at the time. The gelding had a different name then, registered in the station's horse book, before being given his more sinister title.

Lurick was at the Mt Isa rodeo at Calkadoon Park several years later when he heard an announcement for a feature horse called the Barkly Brute. He didn't think anything of it until he met a friend of his, Kevin Ah One. Kevin had come and found Lurick, and was very excited. He and his family used to bring the rodeo horses down as a mob every year from where they were held in a paddock at Carandotta Station. At the time the rodeo had its own horses—cast-offs from stations and drovers, and later horses that the rodeo organisers bred for the job. (These days outside contractors provide the rodeo with its performers.)

Kevin took Lurick to a yard and pointed to a big chestnut horse. Lurick at first didn't recognise it, but when he did he was surprised. 'He had a bit of a buck but he was a quiet horse,' Lurick says. 'Not a kid's pony but quiet and I broke him in.'

He believes that after he left the station the horse was teased and that changed him for the worse. 'I don't know what happened to him after that,' he says. 'I don't know what gave him a fright. I'm not sure how he came to be in the rodeo—the station must have given him to them.'

After Lurick married he took on other jobs that didn't involve horses and settled down to family life. He became a windmill mechanic, fixing the structures that were so vital to outback life, and learned more mechanical skills from an old German chap, so that he could work on tractors and other machinery. Lurick worked for the council in Camooweal and became involved in voluntary work that included driving the school bus and being a firefighter with George Rankine, practising drills with the local constable, Gary Haslem. He also played a big part in helping to revive the open-air picture theatre and was made a life member of the Spinifex Country Music Club for his work raising money for charities. The family later moved to Mt Isa.

Lurick owned a string of handy horses over the years—quarter horses and quarter-horse crosses—and loved competing on the gymkhana circuit with them, enjoying the sprint and barrel races, and the 'pick-up man' event where you would gallop along and pick up your partner, who was standing on the ground. Their children would scream with fright when they watched him compete, recalls Maxine.

It was when he was a family man that Lurick came by the horse that perhaps made the biggest impression on him. It was a bluey-grey foal called Piccolo. Piccolo was born to a mare the Sowdens owned but it was soon obvious that he'd been born blind.

Lurick called in the vet, who said the foal might've caught an infection from his mother, and left Lurick with needles to administer morning and night. At night Lurick put the foal in a crate he had in his truck, fed him grass and milk from his mother, and gave him water. During the day Lurick would put a rope around Piccolo's neck and take him for walks around the yard. Left on his own the foal would have walked into fences or trees or tripped over. The three girls spent time with him, too.

'He was a special horse,' says Maxine, 'like a baby.'

After four or five weeks Piccolo came good, his eyesight restored. Lurick later broke him in and enjoyed riding him, though at a gentler pace than he rode in his youth.

His early handling made Piccolo tamer than he would otherwise have been so when the local priest wanted a horse to come into church for a Palm Sunday service, Piccolo was it. He was dressed in a blanket and led down the aisle, past the pews to the front of the church, interested only in the children and the apples they brought for him.

Piccolo went missing one day. 'He'd gotten out of the yard through the gate,' says Lurick. 'We saw that the gate was open and went looking for him. Believe it or not we found him inside the house at the hostel we were managing at the time.'

But the second time he escaped turned out tragically.

The Sowdens were living in Mt Isa and kept Piccolo, who was about seven, with other horses on the town common where the railway line ran through. He got out onto the track and was hit by a train and killed.

Lurick is now 70 and is perhaps better known for his community work than his early time with horses. He doesn't ride anymore, but if he were to look back on his days with horses, it's not breaking in the 'cyclone from the chute' that he'd remember but a foal called Piccolo.

10

MUSTERING ON

It was a wonderful time, says 73-year-old Bill Batty. Droving cattle all day through the desert plains along the Birdsville Track. Just the riders, the cattle raising dust as they plodded and the far horizons. There's just no better way of seeing the country than on horseback, he reckons. Bill's old enough to have been a drover at the time when the last cattle drives took place on the Track in the 1970s but the one he enthuses about was a re-creation, a chance for all-comers to experience the romance and tradition of droving life—without some of its discomforts.

Bill and his daughter Debbie went on the Great Australian Outback Cattle Drive in 2006, a five-day trek moving nearly 600 head of cattle from Birdsville in southern Queensland to Maree in South Australia. It had all the elements of the old drives: the long hours in the saddle, the stockwhips, Akubras aplenty (though the paying customers wore helmets), campfires at night and the ever-present dust. But it also had twin-share tents with floors and good mattresses, a bar, and semitrailers with hot showers. Riders were advised to wear pantyhose under their jeans for extra comfort—something Bill never did when he was a younger man and rode horses as part of his daily working life.

Bill, a blue-eyed gentlemanly chap, went on the outback ride about the same time that he retired his own horse and put away the saddle. Horses had become redundant to farm life by then but he counts himself as lucky that he was raised, and worked with horses, at a time when they were useful to farm life and when you *had* to ride them.

He grew up in Glen Valley in Victoria's high country on a property called Bogong Views, a homestead standing against a backdrop of meadows and lightly timbered slopes beneath Mt Bogong. God's own country. It was the 1940s, the tail end of the days of buggies. His family owned a car but horses were still used for carting wood or moving hay, shifting logs or herding cattle. The family had at least six in work at any one time so that some could be spelled in between, as well as mares and foals. There was a Clydesdale for the heavy moving work, thoroughbred crosses or stock horses for the cattle work, and ponies for the children to ride when they were helping. But the days when vehicles would take over most of this work were closing in. Bill's father's 1928 Chevy would be joined by a six-cylinder Buick and later a one-tonne utility—the first farm vehicle—then a Ferguson tractor.

One of nine children and the eldest of three boys, Bill was expected to take on farm work from a young age. He recalls always being with his father, watching what he did when he was too young to go to school then helping him as he got older. He watched as the family's Clydesdale pulled the 'reap and binder', the cutter in front slicing the oaten hay, then as the hay fell back on the canvas and was formed into a sheaf and bound by string, to be carted on a long wooden sled to sheds where it was stored. He was with his father as he manoeuvred the Clydie and the old dray to shift wood or take hay to where the cattle were. The Clydesdale was also used to drag logs from peppermint, snow or manna gum trees that had been felled for land clearing.

Bill learned to ride through a combination of instinct and instruction, at first double-dinking on the rump of his father's horse, then riding a pony his father gave him. 'Keep your hands on his withers, son, don't hold them up too high' or 'Always keep the toe of your boot in the stirrup when you go to get off, not your whole foot in case the horse takes off and drags you with him'. Bill rode Nelse—named after Mount Nelse nearby—a stock pony who was a bit of a rogue, occasionally pitching him off with a pigroot or buck. By six he was on his first ride droving cattle.

His father, Bill senior, and uncle, Jack, took several hundred head from Bogong Views, which the brothers jointly farmed, to their lease on the high plains, perhaps half a day's ride. Bill learned how to keep the cattle together by working out wide around them, moving steadily on his horse to keep them calm; too fast and they'd 'break' and scatter. The riders would take the cattle, at their peak 800 Herefords and Hereford–Shorthorn crosses, to the alpine country around Batty's Hut in November then muster them and bring them back down in April or May the following year. They'd return at intervals to check that the cattle hadn't strayed and to 'salt them', replenishing the salt licks that kept them there. The men would call out 'Salt-o, Salt-ooo' and the cattle would come from miles away. The closer cows would smell the salt first and start appearing. The high country is deficient in the salt that cattle need, Bill explains.

The two brothers had a lease for the alpine land, which was national park, and one for the land below it that was state forest. The high plains were more open then, before the government restrictions on grazing, more park-like, Bill says. The cattlemen also carried out fuel reduction burns to reduce the undergrowth and keep the grass sweet. The names of the families that used to have leases there defined the district's spirit and history; the Fitzgeralds from Shannonvale, Kellys from Omeo Valley, Faithfulls and McNamaras from Omeo, and the Maddisons, Wally Ryder and Jack Roper, all from Tawonga. The huts named after them stood as sentinels to their place in history although many of them, including Batty's Hut, burnt down in the 2003 fires that swept through the area.

The families were known for the riders among them and by their horses. Bill senior was regarded as a competent horseman at a time when such a reputation counted, and as a top bushman, too. He rode a piebald called Peter Pan in places where others feared to tread. He was a 'natural' at breaking and training horses, treating each horse as an individual. A photo of Bill senior in his son's lounge room shows a lantern-jawed man, strong of feature, dressed for business at the saleyards in shirt, tie and a Drizabone and Akubra.

Bill senior and Jack's abilities as horsemen were called on at the time of what became known as the 1936 Bogong Disaster: the search for two missing skiers in August that year. It's a story Bill remembers hearing his father relive when visitors called in, particularly those who'd been in the area when the drama unfolded. It all started when a skier appeared at the homestead on the morning of Monday 17 August, dishevelled, frostbitten and exhausted. As Jack made him comfortable over a warming cup of tea, the man told him that he'd been skiing with two others when the weather turned bad and they lost visibility. The three men had been attempting to cross the Bogong High Plains before climbing Mt Bogong when they were caught in a blizzard and got lost. After trying to find a way down, two of them had sheltered in a large hollow log while the third and fittest, Howard Michell, walked down to Glen Valley for help. Jack knew exactly where the log the skier described was and he and his brother saddled up and set out, notifying others in the town of the emergency. A search party that included most of the men of the district, including the miners from the Glen Valley mines, set off on foot. Bill senior and Jack were the only horsemen involved.

Three days elapsed between the time Howard Michell arrived at Bogong Views and when one of the search parties, including

Bill Batty, located the missing skiers—Mick Hull and Cleve Cole. Cleve, an adventurer who helped open up the area for future skiers, was brought down unconscious to Glen Valley but died as medical staff tried to resuscitate him and locals massaged him that night. The story made headlines around Australia and the Battys and the townsfolk were lauded for the way they dropped whatever they were doing at the time to go out searching in the harsh conditions for two days.

Bill senior and Jack's days in the high plains were numbered from 1952 when the state, then later the federal, governments restricted the number of cattle that could be grazed there due to concerns about the effect they were having on the alpine environment. The brothers' herds halved. Bill senior bought some land and started his own farm at Swifts Creek, moving his family there, while Jack remained at Bogong Views. Bill and his sons drove cattle back and forth between Swifts Creek and Glen Valley, a three-day round trip, which Bill junior continued until the 1980s when the family started trucking them there instead.

Bill inherited Bogong Views after his uncle died but lives on the farm at Swifts Creek. He feels fortunate that he saw out his times riding on what he says was a remarkable horse. Lady was a buckskin stock pony, a pretty, feisty mare with a big heart. Bill bought her as a seven-year-old in 1994, after she'd been used in endurance races then as a stock horse. Lady walked out boldly and quickly, and knew her stuff. She was a great traveller who took everything in her stride. You'd just step on her, relax and she'd do the rest, says Bill. Sure, she could be temperamental at times, giving you a boot or a nip if you weren't looking or stamping a hoof on the ground to let you know she wasn't happy about something, but once you were on her, she'd just keep going and going for you.

Bill retired Lady in 2006, around the time the high country leases finally ended, when she was 25. She had a 'bit more in her' but was starting to trip. He let her go at Glen Valley, keeping an eye out for her whenever he was there. She'd always come up to the fence and Bill would have a talk to her.

Then one day in September 2012 a neighbour rang Bill to say that Lady had died. She looked like she'd dropped down walking along and was sitting on her haunches as if she were taking a rest. Bill had only seen Lady a few days before and she seemed to be fine so he was shocked. He pauses. Yes, he was 'a little upset' but, no, he didn't go so far as to bury her. She'd be a skeleton by now, the odd wild dog would've feasted on her, he says. 'Yeah, I miss her.'

He'd stopped riding on the farm, although they still had two remaining horses, because using horses had become impractical and time-consuming. You could get in the ute and drive from point A to point B and back again in the time it'd take you to saddle up and get riding, he quips.

He misses working with horses at home, though. Misses the carefree way you could sit on a horse and take in the countryside while the horse picked its path. When you're driving a ute, you're looking out for stumps or holes—a horse does that for you. He might have stopped working with horses but Bill Batty still rides, thankful for opportunities like the outback horse muster—even if the horse they gave him wasn't as good a walker as Lady.

PART III

Time of their lives

11

THE MAN WHO SOARED HIGH

Maurice Tapp remembers 1948 fondly. Remembers it as the year he was named dux of grade nine at school and for the day he rode his pony Pepper down the main street of Hobart. Being made dux made him proud but the ride was even more exciting. Maurice was fourteen at the time and part of a group of riders who were promoting the Australian Buck Jumping Championships in Hobart that year. They walked their horses down Elizabeth Street and around the block carrying banners advertising the event, accompanied by a showgirl in western kit. There was a fair crowd watching them, Maurice recalls. Traffic stopped. People clapped. Better still, he was able to meet some of the big-name riders who were performing in the event.

Maurice's father had arranged for him to take part in the ride that day. Leslie Basil Tapp was what was known as a 'forwarding agent' for travelling shows, spreading the word about an event coming to town with pamphlets he'd deliver, wedging them in front gates or putting them on the steps of houses or under the door of shopfronts. It was a time when the shows—buckjumping, rodeos, circuses and country-and-western singing troupes—were eagerly anticipated events, particularly in country towns.

Les Tapp knew a number of people who worked in rodeos and in Wirths and Perry Brothers circuses and sometimes worked for them as a clown or a mime artist dressed in black and white, silently acting out the ringmaster's words. Young Maurice loved watching the circus horses in their fancy bridles and wide leather girths; the teams that would gallop around the ring three abreast then change positions; the trick ponies; the horse that could 'count'; and the horse that would pirouette, spin around on its haunches and dance.

He had been interested in racehorses, too, ever since he was a primary-schooler walking past the Elwick racecourse on the way to class, hoping to catch a glimpse of the horses as they trained. When Maurice was in his teens the Tapp family owned a boarding house in Hobart where several jockeys stayed and he would listen to them talk of races and horses, entranced. It was through these jockeys that he was given the chance to ride a racehorse called Peter, son of the champion Peter Pan, which won the Melbourne Cup in 1932 and 1934. Peter was an impressive chestnut with a beautiful temperament and had come a close second in the 1944 Melbourne Cup. He was a bit aged—about nine—by the time Maurice rode him but just being able to gallop him on the track was a thrill.

Les Tapp knew a racehorse owner called Herbert Sylvester Cook and Mr Cook, as Maurice still refers to him, gave him a part-time job at his stables at the racecourse in his final year of school. The stables were run with the same degree of formality and respect as other parts of society were in those days. You always made sure you addressed the owners as 'Sir' and the trainers as 'Mister'. Mr Cook was a quietly spoken, well-respected man who operated a thriving bus company but still managed to train his horses himself. He took Maurice under his wing. At the races on Saturday he explained the finer points of racing and discussed with Maurice how the horses were going. Maurice's duties at the stables included cleaning the stalls, track work and grooming.

In the morning he would change the horses' water and straw bedding and muck out the stables, shovelling the manure into chaff bags then dropping it into a big concrete box. This was emptied twice a week by a man with a pitchfork and taken for use in market gardens. He'd line the stables with fresh straw then let them air out as he took the horses out for their daily exercise. He then ran home, washed, changed into his school uniform and went to school for the day, returning to muck out the stables in the afternoon on his way home. He cleaned the stables on Saturday mornings, 'polished up' the horses for race day, helped lead them to the course if they weren't floated and sometimes stood with them in their stalls before they ran. You always had someone standing with a horse before its race to prevent any interference. Maurice worked on Sunday mornings, then had until Tuesday morning off, and was paid 1 pound 5 shillings for the week's work.

Maurice was offered a job when he finished school as a probationary jockey at Symmons Plains, a sheep station at Perth in the middle of Tasmania. The station was owned by Boyce Youl, who raised Aberdeen Angus cattle, and had a big track for training his half dozen or so racehorses. The probation period was six months—if it worked on both sides Maurice would be taken on as an apprentice jockey. Again, Maurice was liked by the owner and treated well. 'You're a good young man,' Mr Youl would tell him. Maurice helped take the horses to the course at Mowbray, where he'd watch the races through the day and dream of being a jockey himself.

Maurice did race once when he was invited to ride in the Oatlands Cup by a trainer called Archie Oxley. His mount was called John Law, an 18-hand 'gangler' with hooves like dinner plates. He didn't get placed but that didn't diminish his eagerness and interest in the sport. The highlight of any trip to Melbourne was always a visit to the museum to look at Phar Lap in his glass case. Maurice would walk around the grand chestnut to look at him from every angle, wander off to see the rest of the museum, then return to stand in front of the horse again, transfixed.

Maurice's probation training went well. Towards the end of the six months trial Mr Youl asked him if he'd like to travel to Scotland to help bring back some prize bulls. He readily agreed, excited at the prospect of the trip abroad. About the same time, a jockey he'd met from New Caledonia invited him to go back with him to the island and ride for a well-to-do owner there. Maurice said he'd love to go. His father, though, was not long out of the army and had bought a farm as part of the Soldier Settlement scheme. He asked Maurice to leave Symmons Plains to work on it with him.

'What could I do—I couldn't knock my father back,' he says. 'You always tried to help your family—always.'

Yet 60 or so years later you can still hear the disappointment in his voice. The trip overseas, the chance to ride racehorses on the Pacific island or to become a jockey, ended that day. Not long after that he grew too tall to be a jockey.

Days on the farm in northern Tasmania were full, stretching fourteen hours: milking and washing down the dairy twice a day, clearing trees to create more pasture and working in the fields. The dairy produced cream that was separated from the milk and sent to Duck River Butter Factory. In between milking, Maurice ploughed and harrowed an 11-acre paddock ready for sowing barley: four horses abreast with two double sets of harrows. The team was composed of three 18-hand mares—Kit, Bell and Bonnie—and Harry the gelding, who at a mere 16 hands had to push hard to keep up with them. Maurice would also take the team of heavy draught horses into the bush to snig logs to be used to build a Dutch barn. The four horses worked 'inline', pulling the logs in single file through the bush. Maurice would sit on Kit bareback on the harness with his feet on the trace chain.

On Saturday nights, at the end of his week's work, Maurice would stand to attention in front of his father in the lounge room and ask, 'Can I have my wages please, Dad?' His father paid him 1 pound for six-and-a-half days work—but only if the 'cream cheque' had arrived, otherwise it was 'Sorry, you'll have to wait'.

His father was strict in other ways, too. One Saturday night when he was sixteen Maurice was walking to the local dance hall with his brother Brendan when he was flung forward by a kick in the pants. 'I thought I'd broken my back.' It was the local copper, a friend of his father's.

'What was that for?' Maurice asked.

'Your father said to behave yourself,' the policeman said.

That's how it was in those days, Maurice says. 'Straight, clean cut, no swearing. That policeman was a good man.'

But theirs was also a loving family. Maurice recalls being asked to move the ute by his father on his sixteenth birthday and finding a Smithfield cattle dog in there—his birthday present. 'That made me day. I gave Dad a bit of a hug.' He called the heeler with the coat like a patchwork quilt 'Bluey'. Bluey was so smart he could 'count' and 'talk'—he'd tell you by barking when there was a missing cow.

There were other enjoyable occasions that punctuated the weeks and months of hard work, such as the local dances or the arrival of a travelling show. Buddy Williams the country-and-western singer was a friend of the family and came to town with his travelling show. Maurice would go to the concert at the town hall with his cousin, Eddie Tapp, who became a well-known singer himself. The family would invite Buddy over for a big, home-cooked meal afterwards and to 'stop' with them if he liked.

Maurice's father sold the farm and moved the family to Hobart while he realised his dream to build what was billed as the biggest ferris wheel in the southern hemisphere. The family travelled with the ferris wheel for a while before selling it and moving to warmer climes in Queensland so his father could gain some relief from the painful arthritis he had endured for some time. They leased a 350-acre farm at Acacia Ridge, on the outskirts of Brisbane, bringing up all their horses, dairy farm equipment and plant from Tasmania. The weather was 8 or 9 degrees warmer on average but, unlike the rich pasture in Tasmania, the land was maiden bush that carried much less stock. Maurice, his father and Brendan set about making the farm productive. They were always hard workers, the three of them.

On the weekend, Maurice and Brendan took time off to go to pony club in nearby Rocklea, riding horses their father had given them. It was the start of a successful career in showjumping for Maurice. One of his horses, Smokey, proved a good all-rounder at pony club, competing strongly in pony races and novelty events like barrel races, but excelling at jumping. Smokey was a beautiful looking animal, Maurice recalls, dark brown, part-draught horse, part-brumby, only 14.2 hands high but she looked like a thoroughbred. A lively three-year-old when Maurice got her, she loved to jump. You could tell because whenever you arrived at a jumps course her ears would go forward, he says.

The whole family would go to the shows and gymkhanas, taking a picnic and making a day of it. Maurice's father loved competing and watching his sons ride. The Tapp family became so well known for their riding ability that when Maurice drove the horse float into the grounds, other competitors would groan and say, 'They're here again'.

Maurice and Smokey were a formidable pair. They won so many blue ribbons that his mother stitched them together and made a cooling-down rug for Smokey. 'It looked lovely.'

They competed at shows and gymkhanas around Brisbane in the 1950s in an event called 'touch and out' in which competitors would ride over a series of hurdles, as many as they could. The hurdles stood at about 3 feet 8 eight inches high, were made of big, heavy timber planks that didn't fall down when you hit them, and were arranged in a circle. Competitors kept going around earning points until they clipped a jump or the horse refused. Maurice and Smokey once competed in this event in the Brisbane Show and cleared 36 jumps before Smokey, exhausted, crashed and went down at the 37th. They won by ten jumps.

In another memorable event, Maurice was asked to ride a horse over a jump nearly 7 feet high at a performance of the Noel Georges Travelling Rodeo. Noel was a friend of his father's, almost six-foot-four, thin as a whip and a master horseman. Maurice used to love watching his show. Noel had a famous horse called The Inkler, a champion high-jumper who could clear 8 feet. As the grand finale of each night's show, Noel would gallop the horse into the marquee then clear a vertical jump 6 feet 10 inches high. Because of the size of the marquee, horse and rider had to jump then turn sideways mid-air to avoid hitting the end of the enclosure 10 or 12 feet away when they landed. The Inkler was a thoroughbred cross, a 'big bay fella'. People came from miles around to watch his huge jump. But Noel used to drink a bit and one night when he knew he wasn't going to be able to shape up for the show he asked Les Tapp if he could borrow his son to stand in for him.

Maurice, then about 24, was honoured but apprehensive—the highest he'd jumped before was perhaps 5 feet, and all eyes would be on him. The only instruction Noel gave him was, 'Hang on!' Maurice donned his best R.M. Williams jacket over his jeans, and waited nervously outside the marquee for the signal to go. The crowd hushed. A clamour of circus music blasted out of the loudspeakers, the ringmaster made the announcement, Maurice urged The Inkler into a short, swift gallop down the aisle and flew over the jump. While some people stick to a horse like glue when they jump, Maurice used to take the weight off his mount by opening his legs and leaning on the stirrup irons and leathers, rising mid-air as the horse did. It must have been quite a sight. 'It's something I'll never forget doing. It was unbelievable, I felt so privileged!'

He joined his father in another crowd-pleaser at the World Buck Jumping Championships in Brisbane one year. Les Tapp, Maurice and his cousin Dale all dressed in clown outfits and staged a 'pony race'. Les and Dale stood poised for the start of the race next to their ponies—one 14 hands, the next one down 11 hands—with Maurice standing beside a very small Shetland. The compere lined the three of them up ready to race then fired the starter's gun. As Les and Dale jumped on their ponies and took off, Maurice picked up his pony, lifted it onto his shoulders and ran down to the line with it. The crowd roared with laughter.

Maurice's skill jumping horses drew admiration elsewhere. He was competing on Smokey at the Rocklea Show in 1954 when selectors for the equestrian team for the Melbourne Olympics in 1956 saw him ride, approached the judges and said, 'We'd like to see more of this fellow.' Maurice was invited to go to the final qualifying trials for the team in Melbourne but, again, fate and family intervened—he had married recently and needed to return to Tasmania for the birth of his son. 'I couldn't leave my new wife and new son by themselves in Tasmania.' Maurice had met Denice, his first wife, at an end-of-year ball following a year of national service training with the army in 1953. He stayed in Tasmania for a while with his new family before returning with them to Acacia Ridge, where they lived with his parents in the farmhouse.

Maurice kept riding, competing and winning until his father gave up the lease on the farm at Acacia Ridge in 1956 and the family moved to western Queensland to work on building sites. This meant leaving Smokey and the other horses behind on the farm. It was a huge disappointment to Maurice. Smokey was not only a brilliant jumper but had a lovely temperament. She was loyal and a 'real smooch'. The pair had developed a close bond. It was the end of Smokey's competing days, too; she was put out to pasture as a broodmare. She lived on the farm until she died at 37.

While his horseriding days ended when his family left Acacia Ridge, Maurice went on to swap the thrill of jumping for that of speed—racing cars at the Richmond Speedway. Again, it was a family affair. He and Brendan were the joint proprietors of the speedway in the late 60s and early 70s. Maurice drove two custom-line Fords, a GT Falcon and a Morris Mini Cooper S, for which he still holds a track record.

In a life full of ups and downs, Maurice lost most of his possessions, including many of the photos of his horses, in the Ash Wednesday bushfires when he was living at Macedon in Victoria in February 1983. Two years later he was badly injured working for a car manufacturer when the chassis of a truck fell on him. Then 51, he was told he'd never work again. He's still in pain and on medication.

Maurice Tapp might have given up performing in front of crowds a while back but he's still noticeable in public, a dapper dresser with a handlebar moustache. He counts himself as lucky to be married to his third wife, Margaret, and as having a life so rich in experiences.

12

A BOYHOOD OF HORSES

'What has thirty cylinders, two horsepower and flies?'
Answer: 'The night cart.'

It's an oldie but a goodie. By oldie we're talking about a joke that was around before the 1960s when the night cart still rattled along the streets and laneways of Australian suburbs as its driver collected the waste from backyard dunnies. The night cart man would remove the pan from the dunnies or 'thunder boxes' usually located on the back fence, tipping its contents into cylinders on his cart. He and his business was often the butt of such jokes—and worse.

The night cart was just one of the horsedrawn vehicles that fascinated Alan Robertson, who tells the joke, as a boy growing up in the Melbourne suburb of Cheltenham in the 1930s and early 40s. Alan, a vivacious man with a sparkle in his blue eyes, tells his stories gesticulating, chuckling and breaking into song every now and then. His days outside school hours were filled with horses from dawn to dusk and sometimes well in the night, too. Watching, handling, riding, driving and racing them with a school mate—all without owning one. 'Horses were our lives. That's what we lived for.'

Alan grew up in a good place for it. Cheltenham, and Mentone next to it, were then on the edge of suburbia. There was a racecourse and training tracks, beaches further south where the racehorses were swum, riding schools nearby, businesses that used horses and carts, and paddocks with horses in them everywhere you looked. Beyond were market gardens, farms, more paddocks and 'nothing country' that stretched out to the Dandenong Ranges.

Alan knocked about with a boy who came to his school in third grade called Rex Kilburn. Rex not only shared his love of horses but was well connected—he knew Ern Barker, the Olympic rider, who lived locally and whose father owned the bakery. This meant not only the chance to watch the superb rider in action at weekends, but free cakes, too. The two boys would go to the bakery straight after school on Thursdays, get the offcuts from the Swiss rolls, eat them, then head off down to Ern Barker's stables and yards.

Alan would get up in the middle of the night to watch the 'milko' with his lorry and, less often, the night cart man. He and Rex once played a practical joke on the dunny can man, fixing a trip wire across the path at the back of old Mrs Stanley's house then running away to hide then watch and listen to the result. The night cart man duly tripped to the ground with the pan, followed by a big crash and much swearing. The next morning Alan was on his paper round and saw where the mess had been. Mrs Stanley was outside at the time. 'Some awful boys put a wire across the path,' she told Alan, 'and the poor man was on his hands and knees cleaning up this terrible mess on the footpath.'

When he was about ten Alan would go out with the baker's driver on his morning deliveries, waking up at five o'clock to get there and help load the bread into the big covered box at the back of the cart. Ah, the smell of that warm bread, he sighs, remembering it. He was paid 'about two bob' (shillings) on Saturdays to be the offside driver, though the horses knew what to do and very little driving was required. The two drivers would sit at the front under a hood, a couple of spare rolls at the ready to hurl at the 'angry dogs' on the way. The driver would go into each home with a basket holding an array of loaves, rolls and buns from which the woman of the house would make her selection. The horses knew the route through Mentone and Parkdale then would trot back to the bakery along the Nepean Highway when they were finished.

The driver for the local butcher shop, who lived just behind the Robertsons, invited Alan out with him, too. They used to take the meat, butchered in the Nepean Highway shop, to farms and market gardens. The housewives would come out and order what they wanted and the butcher would slice the cut off for them. Alan was just a passenger on this route, asking questions all the time.

On weekends he helped out at Docherty's Riding School in Beaumaris and would stay there afterwards on Saturday night 'talking horses' round an open fire, sometimes until three o'clock in the morning. 'I don't know why they didn't tell me to go home!' he says. 'And my poor mother!' His mother told him years later than she worried about him and would stand on the road at night, waiting, then go back to bed as soon as she heard him walking down the road towards home whistling.

Docherty's had 20 or 30 horses that they hired out at weekends, held in a yard, saddled and ready to go. When the rides were over on Sunday night, the horses would be returned to their paddocks further afield, driven as a mob through Beaumaris, down the main street of Cheltenham and across Nepean Highway to their paddocks in Moorabbin. Alan regrets that he never went on one of the weekly droves through the suburbs.

He and Rex hired horses from another local riding school, Finch's, where American soldiers went to ride on weekends during World War II. The pair, by now teenagers, heard that the legendary American jockey Johnny Longden was racing at Caulfield one Saturday and went to the races to watch him. Johnny had a different way of riding to jockeys they had seen before and raced with his knees bent up

Brumby runner Craig Orchard with horses he's just caught head'n'tailed, in Victoria's high country.

Ginger with Parrot the brumby not long after she got him, in her Sydney backyard.

John Stubbs riding Paddy, the brumby he raised as an orphaned foal, with Mitta the packhorse. Mitta was regarded as a hopeless case but came good with a little patience and retraining.

Jeff Hill jumping Rocket, no saddle, no bridle. Rocket gained a reputation for winning events including jumping and barrel races at rodeos across the Northern Territory in the late sixties and early seventies.

Roscoe the Reverser (middle) and Darrah Dark One (right) soon after they were rescued by members of the Outback Heritage Horse Association of Western Australia Inc. in late 2005. PICTURE: OHHAWA INC

Darrah in 2012 with Katherine Waddington from the OHHAWA at Nannup, WA, some time after Katherine bought her from the charity. PICTURE: OHHAWA INC

Terry Fenton, on Toby, radios in during a search for a missing person in Tasmania, 1984.

Lurick Sowden, a gun horse-breaker in the early 1960s, aboard Piccolo, the horse he reared from a foal.

Bradshaw Station cattle entering the Victoria River at Timber Creek, Northern Territory, at low tide. Ian McBean, a drover in the 1950s and '60s, recalls stockmen riding across the river with their feet up high on the saddle for fear of crocodiles.

Allan Andrews and Westaway (left) with Ian Rush on Coal Black after winning the pickup event together at Saxby Waterhole Campdraft and Rodeo on Taldora Station, Julia Creek, mid-northern Queensland, in 1977.

Mates Ian McBean and Lurick Sowden broke in horses together on Barkly Downs Station, Northern Territory, in the early 1960s and still get together and remember those times.

Maurice Tapp and his show-jumper Smokey won so many blue ribbons at gymkhanas around Brisbane in the 1950s that his mother sewed them together in a cooling-down rug for the horse.

Bill Batty on Lady, right, with John Pendergast on Lewis, about to ride from Glen Valley over the Bogong High Plains, February 2002.

Jane Greenman with Noddy, the world's tallest horse, adorable but a challenge to own, practically.

Colin McKenzie and his trusty quarter horse Lippy beside the Western-style buildings he's created, in early 2013.

Champion duo Harry Ball and Abbey waiting for the presentation after winning the open campdraft event at Woodenbong, NSW, in 1963.

Rider instructor Sally Francis with Lucia riding Casey at her Tooradin Park Estate. Sally has made it her life's work teaching others to ride.

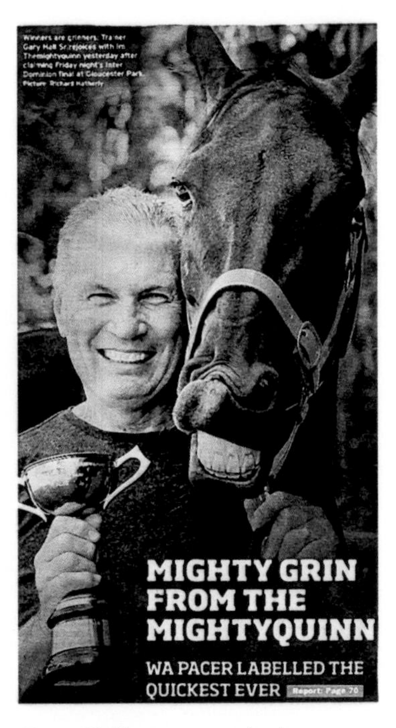

Winners are grinners. Trainer Gary Hall Sr rejoices with Im Themightyquinn yesterday after claiming Friday night's Inter Dominion final at Gloucester Park. Picture: Richard Hatherly

MIGHTY GRIN FROM THE MIGHTYQUINN

WA PACER LABELLED THE QUICKEST EVER Report: Page 70

Gary Hall senior with the phenomenal Im Themightyquinn.

SOURCE: *THE SUNDAY TIMES* 4 MARCH 2012.

It's all in the communication: David O'Meara with Wilbur at the Horse Workshop, Gippsland, which teaches Equine Facilitated Learning (EFL).

high on the horse. Intrigued by this 'American style' of riding, the two friends decided to go to Finch's the next day and hire some horses and show it off. They notched their stirrups high, mounted, perching awkwardly on the riding school horses and rode them to 'The Flats', a sandy, cleared area surrounded by thick bushland in Beaumaris where people met to ride every Sunday. The riding school horses, though, knew all the tricks and when the two boys galloped them onto The Flats the horses turned and bolted for home, ditching their riders. 'That's what we got for showing off!'

Fascinated by racehorses, Alan would get up at four o'clock in the morning, jump on his bike and speed down to the stables in Mentone in time to watch them train. The strappers would take the horses to the beach, riding from their stables along the roads, leading others. The stables belonged to Harry Freedman, and Harry White senior, father of the famous jockey, was the main jockey there.

'Mr White always said to me, "Are you going to be a jockey?" And I'd say, "If I don't grow too tall",' Alan recounts. He used to measure himself against a door handle at home. 'The day I reached that handle I knew I wasn't going to be a jockey. I didn't go back to the stables.'

His hopes of becoming a jockey might have been dashed but young Alan was still a regular attendee at race days at Mentone racetrack (now closed). He, Rex and other neighbourhood kids would ride down on their bikes and wait until just prior to the second last race when the gates opened and you could go in without paying. The boys would walk around and look at the horses in their stables and go to the mounting yard to watch the jockeys get ready for the last race. Other people would come into the course then to pick up the discarded betting stubs, putting them into hessian bags to sort through for any unclaimed place-getting tickets that had been accidentally thrown away. Nicknamed 'emus' because of their pecking motion picking up the tickets, they would spend hours doing this in the hope of scrounging some extra money. These were tough times.

Race-goers would catch Hansome cabs from Cheltenham Station to the course. Alan could see from his home the cabs bobbing up and down Nepean Highway in the dozens, travelling from the city to the station to work on race days. He and his mother would catch the local taxi—Mr Pickering's palomino and cab—from Mentone station home to Jean Street after going shopping in Melbourne for the day, paying sixpence for the trip. The first motorised taxi set up outside the station opposite Mr Pickering's outfit just before the war began in 1939 and people, attracted by the novelty of the car, soon abandoned the horse and cab.

The travelling buckjumping shows were a popular, if less frequent, form of entertainment. The shows would arrive with their rings, marquees and tiered seating, and set up on a paddock on the corner of Nepean Highway and Dandenong Road known as the old 'coffee palace' site. Other travelling shows would also perform there, including circuses. Gypsies moved through the area, too, staying in camps near Mentone.

'Word would get around that the gypsies were in town,' Alan sings. 'While they're selling pegs they're stealing all your eggs.'

The buckjumping shows were just that. 'You'd go to see somebody getting bucked off. We were pretty hickish in those days!' He and Rex would travel the suburbs to see a buckjumping show, trying to crawl under the canvas to avoid the entrance fee if they could. The shows brought their own professional riders and horses, and had a ring and 'a nag' for the locals to try their hand. There was prize money for any local who stayed on the bucking bronco for ten seconds. Alan had a go once as a teenager and rode the leaping horse to a standstill. But when it was time to collect his prize—6 pounds comes to mind—there was an announcement that there would be no prize that night as the horse was 'unwell'. 'It was well enough to buck!'

Buckjumping was all about the horses, says Alan. The riders were irrelevant; after all, they were only on for a few seconds. Aristocrat and Mandrake were a couple of famous horses doing the circuit at the time. Mandrake belonged to the troupe of well-known singer Tex Morton. Alan bursts into song again.

They used to call him Slippery when he came to the show
But they changed his name to Mandrake, there's not a trick that he don't know.
He's thrown the best of riders, the grandest of them all.
And now I'll tell you all about this new outlaw of mine.
So screw down your saddles and make them good and tight.
Back from the ropes, lad, ask him if he is right?
Pick up your mate, lad, he's had a nasty fall
They're all the same to Mandrake, champions and all.

Alan and Rex would go to gymkhanas, too. Gymkhanas were held almost every weekend during the World War II years to raise funds for the war effort, and there seemed to be one on every second weekend in Cheltenham or places a little further afield such as Croydon, Heatherton or Berwick. Alan took his black-and-white kelpie Whisky to one of them and won the Best Tail Wagger event. But the shows were mostly about horses—people were more interested in horses then, he says. Some still had a horse and cart, and kids owned ponies. Horses were cheap—you could buy a pony for 10 or 12 pounds. There were non-stop horse events: the best junior rider, bending races, flag and barrel races and musical chairs. You'd often see steeplechase jockeys competing at gymkhanas when they weren't at the racetrack. Cattle, too, were of more interest then, and people who had a few cows would bring along their best Jersey.

Shows now are far more commercial and less equine, Alan says. They have jumping castles.

Ern Barker was something to behold at the gymkhanas, a real horseman, says Alan, with horses to match. He had the money to be able to buy the mounts he wanted and sell them if they didn't shape up. Alan and Rex used to watch him compete at shows or, if none were on that weekend, they watched him practise at the parklands around the training centre for Catholic sisters, Retreat House, in Cheltenham, where he'd set up his own jumps. Ern was particularly interested in showjumping. In those days that meant half a dozen hurdles spread out around the perimeter of an oval—all 4 feet high, post and rail, with little variation in their appearance. The riders would each go around the jumps flat-out, real 'hillbilly riding', says Alan. Ern used to ride with his feet stretched long in the stirrups in an old-fashioned 'stand up' way before he went overseas in 1955 to gain experience competing before the 1956 Olympics. The Games were staged in Melbourne that year but the equestrian events were held in Stockholm, Sweden, because of quarantine regulations. Ern came back with a new style. 'The Aussies were still rough riding. The Europeans had a far more upmarket style. We were a hundred years behind them!'

Rex used to go to shows with Ern and ride a couple of his horses. His father owned the local timber yards and knew Ern Barker's father. Alan felt he was always the 'kid from up the road' where Ern was concerned but he was often asked to ride other people's mounts in novelty events at shows by those who knew he could ride well. He'd spent his early years as a 'bush kid' in New South Wales and had learned to ride before his family came to Melbourne when he was five. A natural athleticism helped his riding. Alan later became house captain at secondary school and vice-captain of the local football club, though he wasn't as interested in other sports as he was in riding.

The two boys would also 'borrow' paddock horses to ride.

There were any number of horses loose and roaming the area during and after the war—pet horses that weren't being ridden, retired racehorses, unbroken or unhandled ones and aged horses that were so decrepit they looked like they'd almost fall over if you tried to sit on them. Many local men were away during the war and there were less people around to look after the horses. The boys were caught one day borrowing a couple of horses in a paddock, and taken to task by a man holding a shotgun. When the man demanded to know his name, Rex dutifully replied Rex Kilburn; Alan said the first thing that came to his head: 'David Copperfield'. The man didn't believe him.

Alan and Rex used to lure some of the horses into what was called 'Lucas's Paddock', an unused block of land on Nepean Highway, opposite where the huge Southland Shopping Centre is now. A derelict Hawthorn brick house, overgrown with ivy and home to a hobo, attested to the fact that the land wasn't used by its owners. The two boys patched up the old wire fences, coaxed the horses in there with food (or on a bit of rope) and closed the gate. Alan made a bridle out of rope, found an old bit, and modified a little racing pad into a saddle. Worked just as well as anything else, he says. The boys would blindfold a horse while one of them mounted, to quieten it until they were aboard. Some hadn't been broken in. Falling off was second nature.

Alan did, however, hear of the occasional serious, and even fatal, fall. He was shocked to be told in late 1942 that a fifteen-year-old apprentice jockey, Clifford Middlemiss, had fallen off a horse he was exercising in scrub at Beaumaris and was dragged by a stirrup and fatally injured. Jack Crilley, a jockey Alan idolised as a teenager, was killed at Mornington racetrack six years later. Alan used to swim racehorses for one of the trainers at Mordialloc beach so was never far from news about the industry.

The local lads used to have their own races on Sunday afternoon at The Flats in Beaumaris and it was here that Alan Robertson had his most memorable ride ever.

The 'Fastest Horse on the Flat' race was an informal meet, a 'show-off thing' for all-comers held just for the thrill of it. A group of horses and riders—sometimes up to 30 of them—would line up on the track, pelt down it, round a tight bend and pull up at a marker perhaps a kilometre from the start. One Sunday Alan was riding one of his strays, a big black thoroughbred that was obviously used to being handled and to running but was very hard-mouthed. Alan figured that all he had to do was hold the horse back until they rounded the bend then let him go.

They all lined up, horses jostling and toey, and were off, the black horse shooting out from under him. Alan tried to hold him back, pulling hard on the reins, yanking back and thrusting his feet so far forward that they felt like they were somewhere near the thoroughbred's ears. The horse was in front and staying there. Alan pulled harder still as he galloped, sawing on the reins, thinking that if he couldn't stop him they'd keep going and end up at the beach. Soon he rounded the bend and was racing with the horse, knees under his chin. As he turned he saw a huge expanse of water ahead, lying across the track. Realising he couldn't stop the powerful horse he hung on as he took a flying jump, clearing what felt like 30 feet but still landing in the water. The finishing marker was behind them as they bolted towards a pile of timber, felled trees, 6 feet high at least, Alan yanking hard but unable to even slow the thoroughbred, galloping too fast to be able to turn him, either. They flew over the pile of trees. Desperate by now but running out of options he ran the horse into some scrub, crashing around through the ti-trees until he came to a halt, breathless but unscathed.

'Gee, you do some silly things,' says Alan, still incredulous. He wasn't sure how far ahead of the field they were when they won but it was by a fair margin.

Alan took the lathered animal back to Lucas's Paddock and rode him several times after that, though it was hard work because the horse always wanted to go. But it was worth it. 'I felt so important up on his back, this magnificent creature.'

Then one day, Alan unexpectedly got a phone call from the police saying they'd been looking for him—and the horse. An 'old guy' had reported his pet horse was missing and Alan had been seen with it. It wasn't hard to spot the big black thoroughbred, either; it was a former racehorse of some note, a Derby winner, the policeman said. Alan and Rex returned him to his rightful paddock.

Alan Robertson did get to own his own horse. In fact, he has owned many since his horse-mad boyhood and at 81 still has a small herd—though none of them has ever been as speedy as the big black thoroughbred that tore through the flats of Beaumaris that day.

13

THE ACCIDENTAL DROVER

John Steuten came to Australia from Holland in June 1953, a 22-year-old who'd grown up on a small farm that raised chickens, pigs, a few fat cattle and some dairy cows. His family had been farmers for generations, working in traditional ways on cosy, productive acreages, never travelling far from where they lived. In the years after World War II, though, there were few opportunities for young men like John to buy farms, and people were on the move, so he decided to look further afield. He was courting a young woman called Leny at the time and when her family chose to go to Australia, John did, too.

Before he left his homeland, and on the six-week boat trip out, he imagined working in Australia, with vague ideas about agriculture, hoping eventually to work in a dairy as his family had done in Holland. He certainly didn't—or couldn't—have imagined that he'd have a job riding a horse all day in the outback. In fact, he couldn't have imagined any of what he was to encounter in the eight months after he arrived.

Leny's family came to Australia under a scheme run by the Dutch Catholic church, which placed migrants with people looking for workers throughout the country. John, Leny and her sister Nelly and brother-in-law Gerard went to Emerald, in central Queensland, where the two men were employed on a cattle station and the two sisters worked in the small town itself. For John, it was the start of a time of enormous change, adventure and constant surprises.

He started droving almost the day he arrived at Codenwarra Station. The station manager, Neil McCosker, son of the owner, gave him a stock pony, a 'little grey thing' of about 14.2 hands called Tabby. Back in Holland John had barely ridden before although he was familiar with the Belgian draught horses that could be found on every farm and town. 'We only had draught horses when we were kids. If you ever rode any of those horses it was to do the work—not fooling around in the paddock,' he says. 'They weren't much fun anyway—they were too slow for starters!'

When John, in his limited English, tried to tell his new boss that he hadn't really ridden horses before, the manager waved aside the protest and said, 'Don't worry, John, he'll show you what to do'.

And sure enough, Tabby did. 'All I had to do was hang on and that pony did everything,' John says.

He was soon asked by the manager to move a herd of several hundred cattle that were gathered around a waterhole through a gate into the next paddock. But as he approached them, the cattle started scattering back to the bush. John pointed Tabby towards a few stray cows, as he'd been instructed to do, and let him fly. 'I had to hang on for dear life but he knew straight away what had to be done.' The nimble stock pony dashed around between the spindly trees pushing out the cattle as John clung on, pulling his legs up onto the saddle to stop himself being swiped off. John was stunned at the stock pony's skill and intelligence. 'That way I learned respect for the horse. They broke me in on him!' he says, laughing.

John took to outback life immediately. Loved the heat, the open river country with its bluegrass plains and ironwood trees, the azure skies, and the long hours riding alone.

His duties sounded simple—boundary riding and moving the sheep and cattle from one place to another around the property to be, according to which animal they were, dipped, dehorned, castrated, sheared or killed, usually by contractors. But the vastness of the station posed challenges to someone more used to a 25-acre farm. Codenwarra comprised two lots of 60 000 acres each, separated by a neighbour's property 10 kilometres away. John and the other stockmen would be asked to go and move a herd from one paddock to another but the 'paddock' might be 10 miles long and the herd might number 500 cattle. All up there were 5000 sheep on Codenwarra and 3000 cattle, mostly Herefords.

John settled into a routine. In the morning he'd set off on Tabby with his sandwiches, billy can and a satchel of tea. At night he'd eat in the manager's home. 'There was as much mutton and beef as you could possibly eat!' he says. During the day he would check to make sure there was water in every paddock and fix any fences that were broken or twisted. Mobs of kangaroos often 'stampeded', damaging the fences as they hurtled into them. 'There were a lot of dingo trappers then—the government paid a bounty for a set of dingo ears—so there was a far bit of shooting going on that panicked the kangaroos.'

He marvelled that you could ride all day without leaving the property—in Holland you would have passed through half a dozen villages—and that you could spend days without seeing a soul, only kangaroos, wombats and the occasional goanna. The manager told John never to run away from a goanna if it was near you. Lie down flat on the ground and the goanna will ignore you; if you stand up it will think you're a tree and want to climb up. Goannas were always looking for a tree, Neil said.

Unfortunately he only warned John about the wild boars *after* John had encountered one.

John was riding Tabby, checking on some cattle at a waterhole, when he saw a group of wild pigs wallowing in the mud around the hole. Usually all you could see of them were their noses and eyes as they buried themselves in the mud, keeping cool, but this day several families of pigs, sows with litters, milled around outside it. John didn't realise there was a boar among them, too. The feral boars with their 15-centimetre tusks were something to be feared. He looked at the piglets, thinking how cute they were and that he'd take one back to the station and show it to Leny later. He jumped off Tabby, picked up the nearest piglet and put it in his saddlebag, forgetting that piglets squeal. The sows had already scattered but as soon as the piglet let out its first squeal the boar suddenly appeared. It was, as John says, a whopper. He thought the boar was just looking at him sitting on the horse until it raced forward and lunged at Tabby's legs with its tusks. He grabbed a stirrup iron and hit the boar on the head, stunning him long enough to get out of the way and gallop out of range. When he got back to the homestead with the piglet, Neil took him aside. 'John, that was the worst thing you could have done. Boars have been known to break horses' legs,' he said.

Admonished, John asked where he could put the piglet and was told to try the engine room that housed the generator under the house. 'I got up the next morning all excited to see how the little piglet was. But it was gone—it had dug a big hole and run off. I learned my lesson—don't interfere with wild pigs!'

He shakes his head as he recounts another early lesson. Part of his job was to get up at dawn and bring the horses from the 'horse paddock' into a yard and give them molasses before the day's work. The horse paddock was 500 acres in size but the horses were hobbled so they couldn't go far. John realised that the task was taking him longer than it should when the boss came out after an hour and asked, 'What's the problem, John?'

John told him he had to get eight horses but could only find six. He'd looked everywhere and couldn't see the others.

The manager said, 'Did you look at the ground, John?' and when John, puzzled, asked him what for, he replied, 'To see where they went.'

It hadn't occurred to John to check the ground for fresh hoof prints. 'The Aboriginal stockmen could go along in the bush at a trot and spot tracks,' he says, still in awe.

He watched the others and learned ways of handling livestock, such as how to gather and move cattle in a big paddock, or that you could find cattle at waterholes together in the middle of the day—any later and they dispersed back into the bush again.

The cattle needed to be moved every six weeks to yards about a kilometre out from the homestead to be dipped, to rid them off ticks. 'I didn't even know ticks existed! They were the size of marbles. Sometimes we saw cattle that we'd missed bringing in and you couldn't see their skin for ticks. If you left them twelve weeks the cow would be as skinny as a runt—the ticks sucked all the blood out of the poor things.'

John's brother-in-law Gerard was horrified when he was showering one day to find a tick attached to one of his testicles. 'John, come and have a look at this,' he said, pointing to the tick. 'Quick, light me up a cigarette.' John obliged and Gerard burnt the tick off—you can't pull a tick out; you have to burn or salt it off.

'Showering' meant standing under a tap attached to the rainwater tank outside the two-room tin shed the two men shared, using a tin dish to wash or shave. John was amazed to find water in that dish iced over early one morning yet by eight o'clock he had to strip off his jumper as the temperature climbed to more than 25 degrees Celsius. Winter in Holland meant grey skies and days that were cold from beginning to end.

There was so much else that was new about the country and its people.

'I was a total greenhorn when I arrived,' says John. 'I'd never seen anything like the station before. When I first went droving the other stockmen looked at me like I was a bit of a jackass because I knew nothing.'

The manager told John later that he'd reminded him of a 'white rabbit'. John explains, 'We had no suntans and had arrived into beautiful sunshine in a place where everyone was more black than white.' The stockmen all had at least one indigenous parent. John was astounded at their skill with livestock, roping and wrestling cattle to the ground in no time at all and breaking in horses, gently. 'Those horses would come in as feral as anything you've ever seen and within two days you'd be sitting on them.'

He was also surprised at the amount of swearing that went on. 'In Holland we swore but here half the sentences were swear words. But the minute a lady came inside the men would be like lambs!'

Leny, also learning about her new country, got in strife once over it. Leny was living with the McCoskers in Emerald, doing the housecleaning for the family, which had nine children, and sharing meals with them. She used to visit John at the station at weekends, though the unmarried couple came from strict Catholic families and they were living with a 'prim and proper' Catholic family so could never be alone in the same room together.

Leny had been to the station for the weekend, driving there and back with Neil McCosker, who always complained about having to open and shut the 'fucking gates' along the way. Leny made herself useful opening the gates for him so that he wouldn't have to get in and out of the car twice every time they got to one. That evening as the family sat down to their meal her boss, the patriarch of the family, Leslie McCosker, asked Leny what she'd learned that day, as he did every day.

'I learned how to shut the fucking gate,' said Leny.

There was a deathly silence around the table.

Leny went red in the face and wondered what she'd done wrong. After the meal the boss called her aside to ask Leny where she'd heard 'that word' before, telling her that it was a very serious swear word. Leny didn't want to reveal that it was his son who had said it, so instead she replied that it was 'one of the drovers'.

John was often surprised by the characters who peopled the outback. He could never understand why the Norwegian man who lived on the station, for example, smelt so much. The man had stowed his way to Australia by hiding in a cargo hold and jumping ship. One day John asked someone about the strange odour that hung in the air of his room. 'It's metho,' he was told. The boss gave the man tea and a whisky in the hope that he wouldn't drink so much of it.

Then there was the priest, who lived between the station owner's house in Emerald and the church. 'Every night about six o'clock he'd find something to do in his garden next to the boss's house and the boss would call him over for a drink, a couple of whiskies before tea. Every night!'

The fencing contractors were a couple to be avoided, although they did a good job of keeping to themselves. The couple had been hired by the McCoskers to erect a 2-metre high dingo fence made of heavy gauge netting in parts of the property. They lived in a tent, kilometres away from the homestead, collecting supplies the manager would transport in his Land Rover and hang in a tree in a sack for them: meat, beer and whatever else they wanted. 'I used to think that meat would be green by the time they found it!' After pay day, once a month, you wouldn't want to go near the tent for the drinking, swearing and fighting that went on, says John. 'They didn't want to know you.' The pair might've been alcoholics but they worked hard and in the heat of the day.

It wasn't unusual for some of the stockmen to drift off after pay day. On the weekend after being paid they'd take a ride into town, spend their money in the pub, and meet someone who'd say there was a job going somewhere and disappear. Queensland was full of 'drifters' then, John found out.

He continued to have small adventures. The manager asked him to take a herd of 40 fat cattle to the station's abattoir one day. The abattoir was in a paddock on the other side of a river 10 miles cross-country from the homestead. But John had never been there before and had no idea where it was situated.

'I'll help you,' said Neil, 'I'll take you across the river, put the cattle across and point you in the right direction.'

Once they'd got the cattle across the river Neil indicated the direction to John. 'It's over there,' he said, pointing southwest. All John could see was open bush, small shrubs and ironwood gums with nothing to distinguish one part of the landscape from the other. 'Just go straight there. You'll see the shed. The butcher will be waiting for you,' Neil said, turning for home. John and Tabby walked off in the direction the manager had pointed. 'And don't forget, your horse has been there before.'

John ambled along—it was important not to push fat cattle too fast or they lost weight before they were slaughtered. He worried as they walked that he'd missed the abattoir in the scrub. The further he went the more he worried, but eventually he saw the sheds that made up the abattoir, and the butcher. The butcher insisted that he have a cuppa, then wait as he killed two cows and help him skin them. John was surprised not only to find himself skinning a freshly killed cow but also as the butcher cut it into several large hunks, opened the wooden doors to the rear cabin of his old T Ford and attached the meat to hooks inside.

'There were so many flies in there that I couldn't see the butcher. I said, "What are you going to do about the flies?"'

'Just watch this,' replied the butcher, taking his hat off and waving it around inside.

'Talk about hygiene!' says John.

Leny, too, remembers being taken aback watching a butcher deftly slice the maggots off some meat.

John learned a lot about horses, which he'd once thought of as 'just animals', in his time at Emerald, mainly from the station manager. Never panic, was the first maxim. Never belt or kick a horse because he won't forget it. And, the horse will treat you like you treat the horse.

Neil would reprimand anyone cursing a horse or blaming it for something that had gone wrong. 'Don't blame the horse—it's the person that owns them,' he'd say. His father, Les, would similarly tell off anyone hurting an animal. If a cow was lagging or lame and a stockman whipped it, he'd say, 'How would you like to walk ten miles with someone behind you belting you with a whip?'

John was disappointed that Tabby was rested after three months and he was given one of the older, bigger horses the stockmen called the 'cadavers' to ride. 'The stock ponies were as smooth as silk to ride but a couple of the others were like riding a camel, you were glad to hop off them!'

John and Leny's time in the outback ended eight months after they arrived—they moved south in March the following year. Leny had problems coping with the Queensland heat. They married a month after they moved to Melbourne and had the first of five children nine months later. They now have thirteen grandchildren and three great-grandchildren. Something must have rubbed off as there are a fair few riders among them. The couple, who did go on to own a dairy, look back now on John's droving days and Leny's time working in Emerald as a wonderful adventure and introduction to Australia. They say that not a day has passed that they've thought of going back to Holland. There are days, though, John admits, that he's thought about going back to Emerald and to droving.

PART IV

Riding the highway of life

14

JOURNEY OF A LIFETIME

In early January 1983, eighteen-year-old David Smith set off in a handmade cart from Townsville heading for Melbourne, more than 2500 kilometres away. Behind him was the man who'd taught him all he knew about horses, but had treated the orphaned teenager harshly. Ahead was the open road and unknown adventures. David didn't realise it as he gave his cart horse the 'walk-on' but he had just started out on an 'apprenticeship for life'. The time when he grew up. As he left, he was just glad to be free.

The idea for the marathon journey came about when George, the man David was staying with in Townsville, offered David a horse as payment for some breaking-in work. George promised David a part-Arab colt called Chester that he could break in along with the other horses. But by the time he'd finished working on them, David wanted badly to go home to Melbourne. He'd been in Queensland for almost two years and he wanted to settle down and start work in earnest.

'How will I get Chester home?' he asked George.

'Ride him,' he replied.

They realised, however, as they planned the ride back that David had too much gear to carry in a pack: breaking-in and shoeing equipment, a swag, a .22 to shoot kangaroos for tucker, tinned and dried food, and anything else he needed to live along the way. They thought of a cart but Chester was too fine-boned to pull one. Then they thought of buying a trotter, which at least would be already trained to harness, and went to a stable in Townsville to look for one. The man at the stable showed them a horse for sale at 600 dollars, far too expensive for David, then showed them Curie. Curie was a ten-year-old failed trotter with—David was told later—a heart problem. David bought him for 20 dollars.

It took them eight months between jobs to build the cart. George and David fashioned the body of the vehicle out of some folded sheet metal so that it was like a box trailer, using a couple of car springs to attach it to two motorbike wheels. They welded on shafts made out of metal tubing and hammered in hooks underneath the cart to hang pots and pans. George bought a canopy for it from a yard selling marine gear. He taught David leatherwork to stitch the harness for the cart and so that he could carry out any running repairs on the trip. He'd already taught him shoeing, which was vital—the horses would go through a set of shoes every fortnight or so. Then they had a practice run with David driving horse and cart to the beach at Townsville and camping overnight on the foreshore. Onlookers were fascinated by the youth and his vehicle and a reporter from the *Townsville Gazette* soon appeared to ask David about his trek south.

Not long afterwards, David harnessed up Curie, tied Chester to the side of the cart, whistled up his dog Tip onto the tray and was away. He was yet to get his driver's licence and only had his first beer on the day he left.

David, the horses and the dog fell into a contented routine. They covered about 30 kilometres a day on the coastal highway then pulled over at night, by a creek if David could find one, for an evening meal and to sleep. He lived on kangaroo meat, tinned food, powdered milk, damper and anything else people would hand him on the way. By now he had learned how to start up a campfire quickly. He'd done some work on Minnamoolka cattle station at Mount Garnet, and was taught how to boil up a quart pot of tea by a stockman there. David and 'Old Uey', an Indigenous tracker, checked the station's fences together, stopping for a half-hour break at lunchtime to eat the roasted meat they'd brought wrapped in tin foil and have a cup of tea. Uey could boil his quart pot 'rain, hail or shine' in that half-hour while David looked across at him amazed as he struggled with his little campfire, drinking his tea cold. He eventually got the hang of it, sometimes adding eucalypt leaves for a bushy flavour. When he was on the road David had his evening meals by the light of a hurricane lamp, slept in the swag under a dome of stars, happy in his own company and that of his animals. Ah, the romance of it all. But reality would soon hit.

As he drove, people who saw the cart would stop to talk to David and ask him to stay on their farm or station. He was often overwhelmed at the generosity of those who would give him a meal or a beer or put him up. He recalls driving the cart past a pub in one town when he was stopped by a man who had seen him approaching. 'Tie your horses up,' the man said. 'We've got a beer for you inside.' David was never a big drinker but was touched that the men at the pub wanted to get to know him and hear about his travels. They wanted to know why he was doing it and ask whether he was raising funds for a charity, but he'd explain that the trip was something he was doing for himself and so that he could keep his horse.

That interest and hospitality recurred throughout his journey. A singer in a country-and-western band playing at another pub welcomed David in and sang him a Slim Dusty song. 'He thought it was fantastic that a young man was coming into town on a horse and carriage.' Another band that played Dolly Parton and Kenny Rogers passed a hat around the drinkers to collect money for him, a true

Australian tradition. The drinkers were taken in by the pioneering nature, the spirit of adventure of it all. Newspapers in towns along the route wrote stories about David and his 'great Australian trek'. He started to pick up work through the people he met, too. He'd stay a while at a station or farm, doing a bit of fencing or shoeing, rounding up cattle with Chester or breaking in a few horses.

David had travelled barely 250 kilometres when the first accident occurred. He was heading for Airlie Beach one evening on the highway just before Proserpine. It was dusk and the landscape around was dimming. David was approaching a bridge over a creek he wanted to cross so that he could set up camp for the night on the bank opposite. He could see headlights coming towards him on the other side of the narrow bridge so pulled Curie up and waited for the car to come through, worried that it was too close for him to get over first. The car went past. But just as they were about to set off over the bridge David's cart was hit from behind. He saw the dark shape of his dog fly over the top of him as he, the horses and cart tumbled down the banks of the creek. David struggled up the embankment and collapsed, shocked, on the road.

What happened next seemed surreal. The driver of the car that had hit him, a doctor, stopped to check on him then drove off. The next thing David remembered was waking up the following morning aching and looking at the timber walls of a cabin, with no idea of where he was or how he'd got there. He remembered the accident and thought about the horses. Tip was on the floor, sore but uninjured.

David had been taken to a small fishing village near Airlie Beach, picked up by some of the villagers, who'd put his horses in a paddock and brought his shattered cart with them on the back of a ute. He was to stay in that village for several months.

The next day he phoned the police to report the accident and found out that the motorist who'd hit him had already called into the police station. David didn't have lights on the back of his cart, the policeman pointed out, and the matter would rest there. David despaired that his trip was over—the cart was 'finished' and he wondered whether the horses would recover enough to be able to go on. He went to see them the next day and was relieved to find that they seemed unhurt, though subdued.

The hamlet he found himself in was a real eye-opener. It was a collection of cabins that had been hand-built with big rough-hewn planks. There was no electricity. The people lived off the land, mainly on goats they killed themselves, goats' milk and prawns. But they weren't hippies. Far from it.

Take Hopper Williams. Hopper was a big, hard man. Like John Wayne. He'd been a pick-up man in rodeos. He was 63 with a 30-year-old wife and two-year-old child. He invited David out mustering a couple of times. During one of the musters he took David and another young man out to catch a bull. He'd hand-reared the bull as a calf from a bucket but it had rickety disease from eating bracken and was acting mad. David stood back with Chester, not wanting to get his horse in any danger. The other rider, 'a kid on a grey horse', went after the bull, which charged at his horse's flank, lifting it up as the youth tried frantically to get it away, kicking at it with his spurs and spurring the bull's eye out. It took a day to subdue it but the bull was quiet as a lamb after that, David says, and blind in one eye.

Hopper told David that same day to get into the stockyard with another rickety bull and herd it through the gates. 'Scared, are ya?' said another man, as David hesitated.

'No,' said David. He jumped up onto the rails, took off his cowboy hat and waved it through the air in front of the gate to the stockyard. The bull charged at it and went through. 'I was a hero then,' David recalls.

Other Queenslanders tried to tease the young Victorian, too.

David remembers John, the 'head honcho' of the fishing village, playing a practical joke on him when he and a few of the villagers were out in a fishing boat, a long way from shore. John tipped David into the sea. 'I was dog-paddling around in the water while they were cacking themselves.' The water was only half a metre deep, as David found out when his hand hit the bottom.

It was an unusual community but the people in that village couldn't do enough for him, he says. The men rebuilt his cart, making new shafts out of ti-trees. In return he helped them with jobs around the village. David lived like a local, ate goats and prawns and drank goats' milk. But it came time to take to the road again.

The next stop was Airlie Beach, a small settlement that was the gateway for tourists to take boats to the Whitsundays. David stayed with a man who owned a riding school over the hills from the beach, taking out people on trail rides for him. It was an interesting experience. 'He was a bit of a hippy,' says David. 'Walking around in the nuddy, living on nuts. I'd never seen a nudist before. He walked around everywhere in the nude, even when girls were looking!' He did wear clothes when he went riding, though.

David was still, by his own admission, a naive youth. He'd been brought up in suburban Melbourne as a Christian, leading a relatively sheltered life. David's mother died when he was four, his father when he was thirteen and his godfather, a man who'd gone to church with his parents, became his guardian. David's still close to Keith. The time away opened his mind to people who were different to those he knew in the suburbs.

First there was George, the man Keith had asked to give David work as a horse-mad sixteen-year-old. George taught David how to work with horses but nothing the teenager did was ever good enough. A 'mouthy sort of bloke', George put him down, abused him verbally, whipped and hit him. David had no family to turn to, no money to get back home and kept his plight hidden from his guardian. It was a dark time. After six months—by the time he was seventeen—David was seriously depressed. The Salvation Army helped and got him admitted to hospital for depression. The police offered to charge George but David, worried about the repercussions, declined the offer. David went back to stay with George, who was threatening to keep Tip and Chester if he left. George arranged the work for him at Minnamoolka Station. When David returned from the station he wanted to go home but was talked into staying. He says that despite all he endured there, George taught him to be a good horseman.

Billy Walpole's another character David won't forget in a hurry. A friend of Hopper Williams, Billy always carried a six-shooter at his side—just in case he saw a dingo. He packed horseshoes in his saddlebag with the nails already in them and if a horse ever needed a shoe he'd have one on in less than five minutes, never bothering to shape it to the hoof as he hammered.

Then there was 'Wes', a local larrikin David stayed with in a small town near Bundaberg. Wes was a good-looking, hard drinking, free-wheeling young dude who always wore a cowboy hat, had no commitments and never worked. The girls loved him. He'd have one woman in bed, another waiting, and didn't care who else was in the room at the time. 'He got me on to vodka and orange juice,' says David. 'I wanted to be like him!'

But David's real companions were his animals. Tip, the black-and-white kelpie cross he'd owned since he was sixteen and working on a dairy farm in Tasmania, was his best mate. When David fronted up at a station looking for work, he'd be told that they wouldn't have his dog on the property. 'It's me *and* the dog,' he'd insist. Curie was a loyal horse that put 'one hundred and twenty per cent' into everything he did. David remembers reaching a long, steep hill before Airlie Beach at the end of the day. Curie was already tired but kept going, staggering, then trying to walk on his knees to get up that hill. 'Poor bugger, I stopped him and we parked at a reserve at the bottom of that hill and slept. The police moved us on the next day,' he says. But he was closer to Chester. Chester was the first horse David owned and he'd broken him in. He was always an easy horse that David could jump on and ride bareback, in a halter, who would lay down in the grass with him or stand still as he taught Tip to jump on him.

David and his team kept travelling south, never knowing who he'd meet or what he'd be doing along the way. He was driving along the stretch from Proserpine to Mackay one day when a couple of teenage girls jumped out of the cabin of a truck and joined him. The novelty of trotting along in a cart soon wore off, though, and the girls, bored, start jogging alongside it for the exercise.

He stayed for a while in a little house on a station near Gladstone, breaking in horses for the young couple whose father owned it. He also broke in a horse for the local copper while he was there and was surprised to be paid with some dog food and a loaf of bread. When he mentioned the payment to a few locals, they just laughed. Yep, that sounds about right, they said.

David made his way down to Bundaberg then Maryborough. He longed to see the outback and planned to leave the coast road and go inland towards Charleville over the Great Dividing Range and pick up the route taken by Burke and Wills down to Melbourne. But, again, fate intervened.

He'd left his gun on the cattle station near Gladstone and had arranged for the station owner to bring it down for him and to meet him at the turn-off to the road inland. As he waited in the cart he let Curie graze by the side of the road, as he often did. He was daydreaming when Curie suddenly spooked. David barely had time to gather up the reins when the horse took off at a gallop; Chester, tied to the side, bolted with him. David yanked back hard on the reins but Curie kept going. He headed into a property, clearing two cattle grids as he went, the cart lurching from side to side. They'd gone perhaps 2 kilometres when the cart hit a tree and David was flung to the ground.

He knew immediately that the trip was over. It was too much to contemplate rebuilding the cart and starting again. Anyway, he'd felt for a while that he'd been gone from home too long; at almost nineteen he wanted to get on with the rest of his life. He'd travelled more than 1100 kilometres.

David had worried, too, about how to raise the money needed to keep the horses in quarantine for five weeks when he reached Brisbane, which was then required by law. He didn't have it and knew he would have to give up the horses. The owner of the station at Gladstone gave David a thousand dollars for Curie and Chester, promising that they'd be looked after and would never leave the property. David trusted him; he knew they were kind to their animals at the station and would give them a good home. He bought a one-way ticket to Melbourne and flew himself and Tip home.

David, now a farrier working mostly with racehorses, looks back on that trip, 30 years ago, as being one of the greatest things he's done in life. 'I became the horseman I'd always dreamed of being,' he says. 'I was a free spirit then: no responsibilities, no mortgage, no debts, no children, no nothing! That sort of passion died out a bit.'

The relationship he had with Chester was one he's never had with a horse again, he says. He's educating a filly he owns after a long gap from breaking-in work and shakes his head thinking about her. 'Chester was so easy. She's got issues!'

He never contacted the station owner to inquire about Chester or Curie. Better to remember them as they were when he owned them and they were all on the road together.

15

THEY CALL HIM TEX

He can be seen most mornings meandering alongside a stretch of highway in South Gippsland: a lone rider in cowboy gear on a quarter horse with a coat that gleams like polished copper. Curious drivers sometimes slow down to look at the man in the Stetson hat, fringed chaps and lamb-chop whiskers or stop to ask if they can take his photo. Tex, as he's known, and his horse Lippy always oblige. Locals smile and wave as they pass.

Tex, or Colin McKenzie, has been riding the highway for the better part of sixteen years now. And even when he's not riding, he's never out of his cowboy gear. People occasionally ask him why he dresses like he does. He's happy to tell you the story of how he came to be a cowboy, in spirit at least, and his philosophies on life and horses. And about the rough ride getting there.

Colin McKenzie, Victorian born and bred, grew up with a fascination with American cowboys. He would go to the Western Picture Theatre in West Brunswick as a boy to see the Hopalong Cassidy serials, then after television came in 1956, he'd watch the black-and-white westerns that screened on Saturday afternoons. He remembers fondly Roy Rogers and his golden palomino Trigger; Gene Autry the 'singing cowboy'; Hopalong (the actor William Boyd) with his white horse Topper; the handsome star Randolph Scott riding tall in the saddle; and, of course, John Wayne. *Lonesome Dove*, about a 5000-mile cattle drive, became his favourite movie. He loved the romance of it all, the idea of being out there riding on your own on the ranges, sleeping under the stars . . .

Colin played 'cowboys and indians' as a boy, revelling in being a cowboy—they always won. He had a cap-gun and leather pouch and, best of all, when he was about eight his mother gave him and his brother a Hopalong Cassidy jumper each. The jumper had a black background with Hopalong in white, it was pretty special. His mother left them not long after that.

His father was an alcoholic and after his mother departed Colin and his two brothers and two sisters were placed in an orphanage in Ballarat. He stayed there, in between stints in other homes, for seven years. At fifteen Colin joined the navy, leaving one institution only to become entrenched in another. He went to Vietnam four times in the seven years he was in the navy before being discharged as 'unsuitable'. He left the navy confused, not knowing what to do or where to go next. 'I didn't know me elbow from me backside!' he says.

Colin worked in 28 jobs in the next five years, and joined a bikie gang. 'Mongrel' was his nickname. But he always felt at odds with the aggressive culture of the gang and when he met some members of God Squad, a Christian bikers' group, he wondered if they offered a better way. He turned up to his first meeting astride a motorbike with a streaker and a character with horns painted on it. Although he'd been to church as a boy, it was only then, at 24, that he found his faith. He has been a committed Christian ever since. But in the fourteen years that he was in God Squad he had the sense that he didn't belong to the motorbike scene, at all. There was still something missing.

It was only when Colin met and befriended some Texan cowboys at a rodeo in Melbourne that it all started to fall into place. In 1981 the Texans—'real' cowboys who lived on the land and worked cattle—invited Colin to stay with them on their ranch in Lamesa on the high plains of west Texas, once Comanche country. He rode with them on the ranch and went to rodeos where the husband competed in the calf roping and his wife in the barrel racing. For the first time ever, he felt like he belonged. For Colin, 'Being a cowboy is what life's all about.' It's about an attitude rather than a job; about what cowboys stand for and the rules they live by—men who are honest, strong, live and work hard, look out for their loved ones, stand straight and true, and look you in the eye.

And a cowboy's gotta have a good horse, he says.

'I just love the nature of horses. The kindness of them. Horses are gracious animals because they let you ride them—that, to me, is something else.'

Colin's experience with horses as a boy was limited to sitting on a dray behind the draught horse at the orphanage, and going to riding schools occasionally as a teenager. He felt awkward at first when he rode the horses in Texas. You can control a motorcycle, he explains. You just need petrol and good balance. But a horse has its own mind, it's not something to be dominated, at least not in Colin McKenzie's opinion.

'I reckon you've got to be a friend to a horse and let it trust you,' he says. 'You have to have compassion, feeling and a kindness for it. You don't want to be a bloomin' bully—you've got to be firm and strict but kind. I've seen horses belted and flogged by people trying to get them into submission and they're bad horses.'

He's also seen horses molly-coddled and believes that doesn't help them, either. Recalls the time he and a friend, Roy, a horse trainer, went to pick up a horse that a girl wanted educated. When they arrived with the float she presented them with an A4 page list of what to do with the horse. 'What blanket to put on at this time of day, what blanket to put on at that time of day, what we should do with the blanket and how to put it on. We got about twenty yards out the gate and Roy screwed it up and threw it out the window,' he says, shaking his head and smiling at the thought of it. 'Horses need to be cared for but they're not us, they're animals.'

Colin kept riding when he came back from that first visit to Texas, borrowing, then owning, a string of horses, each one better than the one before, until he came by Lippy in 1996. He bought Lippy or 'Sugar Badger Doc', of the famous Doc quarter-horse bloodline, as a four-year-old colt from a breeder. When he first saw him in the breeder's stable, the horse's ribs were sticking out and his tongue was hanging out of his mouth—he'd been in an incident when he was tied up and had somehow damaged a nerve in his face, paralysing one side of it. But Colin could see he was a good-looking horse. He bought him for a reduced price and named him Lippy. He rubbed vitamin E cream on his face for several months until the nerve came good, and had him gelded.

In the eighteen years since then Lippy proved to be not just a good horse but a great horse, says Colin. He goes wherever he's asked and isn't bothered by the semitrailers that rumble down the highway—'He doesn't put any fear into me.' He loves a good run but knows when to stop. And he just gets better with age.

'He's so gentle,' says Colin. 'Every child loves horses till they meet one then they're terrified. Lippy's so gentle I encourage kids to pat him. He likes people.'

Every year Colin and Lippy drop in to visit a group of people with dementia who stay at the bed-and-breakfast place on his road. 'They have their photos taken with him and some of them sit on him. These are frail, elderly people. It gives them confidence and brings them out of their shell. You can see it in their eyes. Some of them say, "I remember I used to have horses." It gives them something to hang on to.'

Colin and Lippy have competed in barrel racing, team pegging and 'wild cow milking' competitions, and gone on the Snake Island Muster further up the coast in Gippsland. Lippy had started reining, learning some of the twists and turns of western riding, before Colin bought him but apparently wasn't suited to it, which doesn't bother Colin. 'Just because he's a certain breed doesn't mean that he has to be good at it,' he says. 'He doesn't have to be a champion anything. Horses are there for people to enjoy.

'People ask me, "What do you do with your horse?" I say, "I ride it." It's not about whether you do reining—though I've got nothing against that—or how many shows you've been to or how fast you can go or how much you spend on your horse. It's not about ego, it's about loving things—people get close to animals. It's about fun and what you do for other people. It's about what's inside that counts.'

Colin is content just riding along, enjoying the scenery, singing and whistling, talking to people he meets on the way, letting Lippy stop to have a chew of a plant or giving him a run. He reckons his horse enjoys getting out of the paddock for a walk. He's a good traveller. Bart the heeler used to come with them from the time he was a five-week-old pup, then Doc the blue heeler cross, but Stella the Staffy, the new dog, steers a wide berth around Lippy after her first encounter with a hoof and stays at home.

Colin sometimes stops off at the general store in Kilcunda, tethering Lippy outside. He took him once to the roundabout at the turn-off to Phillip Island, where a mass ride of bikers goes through on their way to the Phillip Island motorcycle grand prix each year. Lippy didn't flinch as hundreds of bikes roared past them barely metres away. It was a hair-raising sight. 'It's good practice for him!' Colin says.

The pair trust each other although Colin says you can never take a horse for granted. He's been kicked by other horses and had his face stitched, been knocked off under trees, booted in the backside coming off at a log, and ended up once with his boot and leg stuck on the top strand of a barbed wire fence after the horse he was riding bolted.

He acquired the name 'Tex' in 2006 after a local magazine, *Coast*, published a story about him and dubbed him that. When people first starting called him Tex, he'd think 'Who?' and is still surprised sometimes to hear it. A bank teller recently finished a transaction—a bill paid into his account by someone else—looked at the reference on the transaction, handed over the money and said, 'Have a good day, Tex.'

'And I didn't even know this lady!' Colin says, laughing. 'But it's a good name, a friendly name.' It was a name that took him further along the road to being a cowboy.

Colin never stops adding to the theme. Over the years he's acquired hundreds of bits and pieces of western, old-time and cowboy paraphernalia that he keeps in sheds he's built with facades that look like they're straight out of the Wild West. The walls are covered with the skulls of horned steers and horses, Texan numberplates, old tack, wagon wheels, you name it. The rear of his big red F100 is covered in cowboy stickers—'Jesus loves cowboys, horses and dogs' says one—and painted with figures of horses and riders. Colin's wife Robyn and the family support his cowboy ways; he can turn up anywhere with them in his Bailey's hat, vest with Texas Ranger badge, bandana with its steer scarf slide, jeans, chaps and cowboy boots.

As for the romance . . . Colin did sleep with his horse one night, in a stable, if not under the stars. He was helping with the judging at a whip-cracking competition at Woolamai racecourse and decided to ride there the afternoon before and stay overnight. (The racecourse pavilion proudly displays a mural Colin, a signwriter, painted on one of its walls.) But spending the night with his horse wasn't quite like the movies. Lippy chewed and chomped and moved around, and 'had a leak', keeping Colin, lying on a deck above him, awake. The mosquitoes were worse. Colin ended up moving into the jockeys' room to sleep. And, he concedes, real cowboys—the young, underpaid cowhands—probably had to eat refried beans and stale bread and drink coffee you could stand a stick in.

Colin bought another horse not so long ago but it turned out to be 'a mongrel of a thing' whose repertoire of bad behaviour included dropping to the ground and rolling over on top of him. He sold it and spent the money on a lawnmower. He's back enjoying the horse he knows and loves. People who see him when he's out somewhere working or in town shopping always ask him, 'Still riding that horse?', which makes him smile. Lippy's still got a fair few miles left in him as Tex rides him on that highway called life.

16

TROTTING BACK FROM THE BRINK

Cornelia Selover wanted a horse for as long as she can remember, even before her mother took her on her first pony ride at the St Kilda foreshore at the age of five. She's not sure where the urge came from; she was a 'city kid' with no connections to the country, growing up in a theatrical family where everyone went to work at five o'clock in the afternoon and no one was particularly interested in horses.

Corn soon started helping out at the ten-cents-a-pop rides, leading the ponies around the sandy track by the bay, just to be near them. If she was lucky Mr Govern, who ran the rides, would let her ride one of the ponies the hundred metres or so to the truck at the end of the day. The rides only increased her fervour to have her own horse and so she did what any horse-mad girl would do: she pestered her parents. She hinted, nagged, cried and threw tantrums, never letting go of the idea.

It didn't work immediately, but when she was ten Corn's mother quietly arranged with a friend of the family to loan her a horse for the summer holidays. The family always went to Point Leo beach on the Mornington Peninsula for the vacations and her mother had it all planned. The friend, Joe, arrived towing a horse float one day, ostensibly to drop off an ice-fridge for them to use in the camp over the summer. He asked Corn if she would mind going to the float and bringing a bucket back from it. She took a look inside the float, got out the bucket and came back, puzzled.

'Why is Goldie in there?' she asked.

Her mother told her the horse was hers for the whole summer. She could stay in the paddock behind the general store. Corn was beside herself.

Goldie was a 15-hand bright chestnut, a gentle thoroughbred with an obliging nature. Corn rode her bareback on the beach, double-dinked her friend Pat and pulled Pat along in the water holding onto Goldie's tail. Pat's father, a farmer, saw her one day when Corn was riding her. 'We could never afford a horse like that,' he said. Corn was upset that he'd said it, and more so that it was true. Even as a girl who desperately wanted a horse, she wished she could have given Goldie to them.

At fifteen Corn finally wore her mother down and she got her first horse. Her mother was teaching ice-skating at a rink in Oakleigh and in the paddock behind the rink was a horse belonging to the daughter of a woman she was teaching. The horse, and all its gear, was for sale. Ready made! Corn had a ride and popped him over a few jumps. 'Mum, he's fabulous!' she cried. They bought him.

Blitz was a solid, cresty-necked bay with a character—they found out—to match his name. Corn couldn't catch him. He'd kick furiously if you tried to pick up his feet, float him or girth him up. The farrier had to tie up his leg to shoe him. He put people in hospital. Corn loved him regardless.

She bought her own horses after she left home, at one time in the mid-80s bringing one of them up from the country where it was agisted to the inner-city suburb of Fitzroy where Corn lived. She kept Sandy the Connemara in a yard at the back of Gertrude Street, opposite where the restaurant Cutler and Co is today. She'd ride from Fitzroy to her mother's house in Brighton on the opposite side of the city, travelling through the Fitzroy Gardens, over the footbridge that leads to the Melbourne Cricket Ground, up the Tan jogging path, through the gardens near the Shrine of Remembrance, through the parklands around Albert Park Lake, then ride along the beach to Brighton. She was stopped once by police, who spotted her cantering happily up a grassy stretch to the Shrine. What did she think she was doing, they wanted to know. At other times the police would remind her not to ride on the footpath, which she enjoyed doing for the views it afforded of the front yards along the way. Sometimes she'd ride towards the beach down busy Flinders Street, on the edge of the CBD, turning a few heads. Sandy was a 'great little horse' and one among a string that Corn has owned in the decades since that she counts as a 'significant horse'.

When she was in her late twenties, Corn decided to move to the country where she'd be closer to her horses. She shifted to Kinglake, a small town northeast of Melbourne sitting near the tail end of the Great Dividing Range. It was a decision that would nearly cost Corn her life.

At first she rented a house on 100 acres. She was singing in bands for a living at the time—she'd trained as an opera singer—and working as a freelance illustrator. She was later employed as senior designer with Treloar Australia, a corporate fashion company. Then, one day in early 1998, an idea took form that would let Corn match her work with her passion for horses: an equine clothing company.

Corn's neighbour Samantha noticed some line drawings she'd been doodling of her niece's fat pony Jellybean. 'That's really cute—let's print it on a T-shirt,' Sam enthused.

Corn, an intense, focused woman with a crinkle-cut blonde bob and a bit of style about her, hated what she calls the 'stretchy beige jodhpur look' and thought it would be fun to create something different in horse clothing. The two women made a small range of T-shirts, at first buying ready-made T-shirts and having them screen-printed, folding and wrapping them in freezer bags on Sam's kitchen table. They called their garments 'Tuk-Tuk 100% Horse' and sold them through markets, horse shows and events in Victoria, then through stores wanting to stock them.

In late 1998 they were approached by a man recruiting stallholders for a new horse exposition called Equitana that was going to be staged in Melbourne in the following November. The two women went to a business meeting held for the exhibitors several months before the expo to inform them about how to make the best of their stalls. They listened, excited but intimidated, as the speaker addressed the gathering using terms like 'maximising exposure' and talking about the mailing lists of the 50 000 people they'd all have. Corn and Sam exchanged glances—they didn't have a mailing list. They wondered if they could cope and tried to back out. The salesman, Fraser, told them they *had* to come.

Their stall at Equitana was swamped—people loved the clothing with the little fat pony logo. An investor who'd been observing the buzz around their 3-by-3-metre stand, and whose wife had bought Tuk-Tuk clothes, approached them asking about their 'plan' for the designs. The two women shrugged their shoulders—they didn't have a plan. On the Sunday night that Equitana ended they took home 13 000 dollars in cash, laid the money out on Samantha's bed, drank champagne and whooped with joy.

They started negotiating with the investor. Having money behind them meant an advertising budget, being able to show off their clothing more, and manufacturing the garments offshore—all aspects of business that Corn knew about from her days working at Treloar. The business grew from the pair working in their own homes to a small shopfront, then a larger shopfront, then a warehouse in Lilydale on the eastern edge of Melbourne. The range of clothing expanded from T-shirts to pants, jackets, tops and accessories, all with versions of the Jellybean logo. The clothes were sold through 200 stores in Australia, as well as to the United Kingdom, America and to any of the company's 5500-strong mailing list. Tuk-Tuk was a great Aussie success story.

In 2000, Corn married Damon, a man she met when they were student pilots involved in a light plane crash together. The plane had been taking off when it had engine failure. The couple bought another property in Kinglake after they married, gradually adding more horses. There was Handsome George, the placid quarter horse, now 38 and retired; Mrs Bailey; her offspring Odile, a powerful Irish sport horse, solid and sure of herself; and Oyster, her full sister; and two other thoroughbreds, Carol and the sensitive and troubled Onion.

Onion changed Corn's attitude to owning horses. She'd bought her after Handsome George because she wanted a horse with more vigour. George was too relaxed. Onion is quite an ugly little thoroughbred, Corn says, the sort of horse that would go straight to the knackers if she came up at a sale. Bay with lousy conformation and bad legs, she'd been bred as a racehorse but had turned out too short and too slow, then was thrashed around for a while before being sold. But Onion had a lot of go in her and that's what Corn wanted.

Corn discovered just how much go when she got the mare home. She let her out in their 85-acre paddock, Onion took off and she couldn't get near her. Onion was terrified of everything, including Corn. If she could round her up and catch her, the horse would bite when being handled and buck when being ridden. She'd panic and take off.

Corn tried every trick she knew but nothing worked with Onion. She rang the wife of Tuk-Tuk's investor, Cindy, whom she thought might be able to help. 'What am I going to do?' she pleaded. 'She's so afraid of me I can't get near her. The only thing I can do is run her into a yard and rope her like a cowboy and I don't want to do that because she's already terrified of me.'

Cindy introduced Corn to a few ideas of natural horsemanship as taught by Pat Parelli. She taught Corn 'the catching game': how to get the horse's attention by rustling a plastic bag, then getting her to take a step towards you, then follow you, until she eventually allowed herself to be haltered. It took three months and a lot of patience but it worked. And that set Corn on a new path in the way she managed horses.

She took Onion to natural horsemanship clinics and found out a lot about herself and the way she handled horses, realising her timing could be better and that actions she made were 'rude' to a horse and liable to make it react in the wrong way. In Onion's case, that meant a monster rodeo buck or kicking you in the ribs. Corn has been to hospital three times in the past ten years after Onion reacted to something she did with her. Now she treats Onion like a highly strung friend, respecting her for what she is, giving her the time she needs rather than forcing her to respond to commands. The patience and 'softness' has paid off. Onion is smart, tough, quick on her feet and never says no. She'll go anywhere and try anything. She's great with cows. People comment on how quiet she is on the streets. But Corn had to win her trust—and still has to—every second of every day she rides her.

Life was full for Corn before the Black Saturday bushfires struck on 7 February 2009. She owned a successful business with a close friend, was married and living on acreage with a beautiful backdrop of mountains and bush, with her eight horses. But all that was to change.

Corn felt instinctively on the morning of the seventh that the day would turn out badly. A nasty northeasterly wind was already fanning the tinder-dry country around them when she awoke. A heatwave the week before and dangerously low humidity had fire authorities warning of extreme conditions for the day. Corn, unsettled, went outside at five o'clock in the morning to move the horses from their yards into the 'big paddock' in front of the house. If fire struck at least the horses would be able to flee from it, she thought. The 85-acre paddock had a long driveway along one edge that Corn and Damon kept mowed on either side as a buffer to the bush beyond. The couple, keenly aware of bushfires because of one that encroached on them two years before, also kept the area around their hilltop home mowed and green. Corn got on the ride-on mower and mowed round the house again that morning—just to make sure. Mowed it until it was almost dirt.

A dozen friends were due to arrive with their horses for a 'play day' of riding and talking, some of them driving with their floats on the road from Whittlesea. Corn rang them early. 'I don't think you should come,' she said.

They protested, saying, 'Oh, it'll be fine, it's just some hot weather.'

'No, don't come,' Corn said. People were killed on that road later that day.

By mid-morning the wind was blowing at more than 125 kilometres per hour. Corn and Damon finished their bushfire preparations—filled the bath with water, got the towels out, woollen clothing ready, made sure the roof gutters were full of water and so forth—and listened to the radio. The power went off. They turned on the radio in the car and listened as broadcasts reported fires breaking out around them: first Kilmore East, then Wallan and Strathewen and Whittlesea. They watched as columns of smoke appeared around them—their house was on a hill and had a 360-degree view.

Gradually Corn realised that the fire was not going to pass by them as it had two years before. Despite all their precautions and vigilance, Corn started to panic. She thought, 'Shit. We're in deep shit. No one is going to come to help us. We're isolated from everyone.' The voice on the radio kept upping the death toll.

Corn could see a massive pyro-cumulus cloud rising thousands of metres into the sky and could see flames flaring in it. She rang her friend in St Andrews, pleading. 'I'm going to be killed. Will you have my horses?'

The smoke arrived before the fire front. Birds started hitting the roof, falling out of the sky, dying. Corn and Damon, standing in the backyard, watched the sky suddenly blacken in front of them. Pitch black, like night. Then, chaos. Somewhere people were screaming. Explosions sounded. They ran searching in the darkness, scrambling for the back door several metres away with not enough time to even take a breath. The firestorm approached with an ear-shattering shriek. They'd just slammed the door shut when the firefront roared like a cyclone over the house—and kept going.

Corn's world collapsed in those moments. As Damon fought to put out the spot fires around the house, she shut down. It's said that there are three responses to fear in the human body—fight, flight or freeze. Corn froze. After being resigned to death, life would not return. As Damon kept putting out spot fires throughout the early evening and into the night, she sat at the window, 'staring like a dog'. At night she slept with the help of some sleeping pills, on a mattress in the garage—the room furthest from the fire. She had seen the fire burning through the windows of the bedroom opposite and could not bear to go in there, let alone feel comfortable enough to sleep. Damon kept working on the spot fires.

The following morning she was still so shocked that it took a while to even think about the horses. She hadn't been worried about them—there were places in the big paddock where they could shelter and she trusted that they would be safe. Odile, her big black mare, had been born 'capable'—she would have led them to safety, Corn thought.

Corn asked Damon if he had seen the horses when he came back after a round of checking for any breakouts of fire. The paddocks were blackened, the bush beyond them razed. But the horses were fine, he said, must have sheltered in the gully that ran through the paddock. Others weren't so lucky, she found out later. More than 40 people died in Kinglake and Kinglake West, hundreds lost their homes. Horses, too, by the hundreds were trapped and burnt. Kinglake was all but obliterated in the Black Saturday fires.

Corn was deeply traumatised in the aftermath of the fires and went into a downward spiral. She felt like she couldn't cope with anything. She tried to give her horses away, couldn't even look at them, or sit with them in the paddock as she used to do. She was unable to work, leaving Tuk-Tuk without its creative force and half its leadership. Her marriage began to break down as the house became more and more tense.

It took months for the post-traumatic stress disorder that had her in its grip to fully strike but by the time it did Corn was suicidal. Samantha, her business partner, friend and the 'rock of my life', was unable to cope with the demands put on her by her troubled friend. Eventually she said, 'I don't want to see you any more' and Corn didn't blame her. She could understand why Sam needed the break.

It was at a time when there seemed to be nothing left in her soul that she started drawing. Corn had studied art and had always drawn. Now she drew dots representing the birth of this life, and another dot the beginning of the next, and connecting lines all that happens in between. Lines that connected symbolically to the next minute, the next day, the next life. Absorbed, she drew books full of lines, and called them 'I wish you were dead'—'you' being whatever it was that was worrying her. The lines, which she would later realise resembled the spiky ridges of hills burnt by fire, were replaced by tree trunks she painted in oil pastel, then paintings of water and reflections, representing transition.

Her art, and the counselling available to people affected by the fires, helped. But it was a horse that set Cornelia on the road to recovery.

By the end of 2010, Corn had suffered the breakdown she knew was coming and underwent treatment for it. She was still disinterested in her horses, unable to feel anything. Then, one day in late 2010, over eighteen months after the fire, Corn felt she had to get out of the house, saddled Onion, pointed her towards the front gate and said, 'Let's go.'

The pair started trotting. They trotted and trotted and kept trotting until the day was done. They trotted the next day and the one after that, all around Kinglake, Pheasant Creek and Kinglake West; it didn't matter where as long as they were moving. Horse and rider found comfort in the rhythm and steady motion. Something in Onion settled as she responded to the routine. Corn took solace in the movement and eventually found a new energy. She remembered what a great little horse Onion was.

They trotted all through summer. Onion learnt to follow the white line on the road, right eye on it, trotting in cruise control, hardly changing her breathing. Eventually Corn trotted to friends' houses to visit then got back on Onion and kept riding. Her world started opening up again.

The legacy of the firestorm went on, however, but Corn was better prepared for it. In mid-2012, Tuk-Tuk was sold to a Western Australian company, Corn and Damon divorced and the Kinglake property was sold. Corn, who'd been living in the Tuk-Tuk warehouse office, decided to take to the road again. This time with Odile, in a float, and with her paintbrushes.

After a couple of stops in Victoria she headed north to a retreat called Banyandah at Howlong just over the border in New South

Wales, the first destination in Corn's loose plan to be an artist-in-residence around Australia. Banyandah is a centre for natural horsemanship and also a gentle, spiritual place. Corn thought she'd stay there for two weeks and ended up staying two-and-a-half months. She sat with her sketchpad by the Murray River, in the filtered light beneath ancient red gums, drawing, losing track of time. She let go of 'I wish you were dead' and began a series called 'White Tree'. She spent time with Odile and remembered what she loved about the self-assured mare.

Other series of paintings followed as Corn and Odile moved south to Victoria's high plains. In 2013 Corn travelled to America, staying on a ranch in Colorado, designing jewellery for a respected metalsmith, then moved on to Herefordshire, England.

As she travels Corn isn't sure where the new phase of life is headed but she's content to be going there, on the road, just following the white line, the way she did when she trotted back from the brink.

PART V

High achievers

17

HARRY'S DREAM

In 1944 a sixteen-year-old boy sat in the grandstand at the Sydney Royal Easter Show and made a vow. Harry Ball was at the show with his mother and sisters watching the campdrafting, thrilled by the speed and skill of the horses and riders that seemed to move as one. But Harry was there to watch one horse in particular, a mare his family had owned only months before. Harry's father had died the year previous after a campdraft accident and the family's horses, including the mare, had been sold. Harry grieved the loss of his father and was heartbroken about the horses. 'I'm going to have a horse one day,' he vowed. 'And I'm going to draft in the Sydney Easter Show.'

Harry did that and more. More than he could ever have dreamed.

Harry grew up in the Lower Macleay, below Kempsey in New South Wales. His father and uncles managed a few properties in the area but when his father died his mother, who didn't come from the land, moved the family to Sydney. Not one for city life, Harry bought a first-class ticket on the train to Tenterfield, in the New England area, right after the Easter Show, hoping to meet graziers returning from it and to get work with them. He met a well-known contract drover on the train, Edgar Carroll, and later worked with him on a few big cattle drives, including one from Roma in Queensland to the Victorian border.

The teenager spent some time working in the New England area, moved further west, and then made his way back to the Macleay in 1949, living near Willawarrin, a small town in the Upper Macleay Valley. The Upper Macleay, inland of the mid-north coast, is a mixture of river plains, pasture and the heavily timbered slopes of the Great Dividing Range. The bush country is steep and rough in places; you had to be a skilled rider to get around it.

A compact man with curly black hair and twinkling blue eyes, Harry was known around the town as good company and a responsible worker. In his early twenties he took up a job as a contract stock hand on Toorooka Station, a beef cattle property close to Willawarrin, one of several properties in the district owned by the Hill family. It wasn't long before a young woman working in the office of Toorooka's butter factory caught Harry's eye. Coral loved dancing and Harry was a good dancer. They went to dances at the Willawarrin hall, played tennis and went to rodeos and on picnics together. Harry and Coral married in 1952. Harry was working for Fred Hill, who owned Toorooka, and became mates with Fred's nephew Theo, who managed properties owned by his father Frank in the Upper Macleay. Harry and Theo spent long hours in each other's company mustering and carrying out cattle work, and shared a passion for good horses. They went to campdrafts together on weekends, travelling up to 200 miles around New South Wales to compete in them. Harry had a reputation as a skilled campdraft rider and was often asked to ride other people's horses at the rodeos as well as his own, but the right horse eluded him.

So he decided to breed his own, a *champion* horse. He studied the bloodlines of the best stock horses around and came up with what he figured would be a winning combination. The Willawarrin publican Hugh Flood owned a mare called Joy's Pal—named after his wife—which Harry had competed on in campdraft events. He agreed to let Harry use her to produce a foal.

Joy's Pal was by Radium II, a successful campdraft horse in the 1930s, who went back in bloodlines to the great Cecil. Cecil, born in 1899 in Glen Rock, New South Wales, was one of the original foundation stock horses, a cornerstone of the Australian Stock Horse Society. He was a brilliant performance horse, winning in campdraft and other equine sports. Cecil was such a formidable competitor that his owner W.H. 'Black Bill' Simpson was asked not to bring him to the 1913 Geary Flat rodeo because the other competitors would be put off entering and if they did enter they'd be thrashed.

Harry chose another talented performer as a sire—Radiant, who was also by Radium II. He'd competed on Radiant, or 'Old Ray' as he was called, for his owner, Roy Harwood. The matching of Joy's Pal and Radiant meant the foal would have an enviable 'double cross' of Radium II in its pedigree. Harry could hardly wait for it to be born.

Abbey arrived on Australia Day 1955 in a small paddock behind the Willawarrin hotel, a jet-black foal with no markings. Harry was excited; now he couldn't wait to break him in. Harry and Theo watched Abbey grow from a wobbly-legged foal to a spirited colt. Harry started training him from a young age, and broke him in when Abbey was well short of two.

A close friend and fellow competitor, Jack Hope, who lived about 10 miles away, recalled years later the disbelief he felt when Harry told him he'd broken Abbey in and was working him with cattle; Abbey wasn't much bigger than a weanling at the time. Harry suggested Jack take him for a ride, pointing to a skittish Angus steer on the flat. 'Work the beast around a bit,' Harry said.

Jack was hesitant, thinking the horse was 'only a baby'. He got on reluctantly but got off astounded. 'It won't be long before you win a campdraft on that horse,' he said, handing back the reins.

Harry did. He won his first campdraft at the Taree show on Abbey when the horse was only 22 months old. He knew then that he had a great horse. Abbey had been born with cattle sense and had shown that ability early. He had plenty of 'go' but learned quickly and handled well.

People started talking about Harry and Abbey—and didn't stop. Abbey kept winning campdrafts and winning them outstandingly. He became known as a tenacious horse that wouldn't let the beast beat him. Harry and Coral travelled to rodeos and shows around the state, enjoying the road trips away together and revelling in the unfolding success of Harry's horse. In less than seven years, Harry and Abbey won 23 campdrafts, including the Duke of Gloucester Cup at the Royal Easter Show in Sydney in 1961 and again in 1964. They came second in the World Championship Campdraft in 1962. But they were more than a competitive team—Harry and Abbey became soulmates, says Theo's wife, Bonnie. They understood each other and Abbey knew exactly what Harry wanted of him. Harry lived for that horse.

In June 1964 life was good for Harry Ball. He and Coral had an 11-year-old daughter, Christine, and became the parents of twins Tracey and Hilton on 4 June. The Balls owned a successful dairy farm at Frederickton. Harry still worked with his friend Theo, and he'd recently won his second Duke of Gloucester trophy with Abbey.

Then, on the night of 23 June Harry was killed in a car accident. He was coming home from work at Toorooka, and was less than 5 kilometres from home when he ran into the back of a semitrailer that had broken down on the road. It was foggy and dark and the truck had no hazard signs. Harry died one of the most famous horsemen in Australia. His funeral cortege at the service on 25 June was three miles long.

In the blur that followed her husband's death, Coral Ball was approached by people from everywhere wanting to buy Abbey, all saying that Harry had promised to sell the horse to them. She knew that wasn't true. She wouldn't sell him. Harry had loved the horse and she was worried that another owner would exploit him. More than this, she didn't want anyone else to ride Abbey; Harry and Abbey had been a team and Harry had got the best out of him. But Abbey became too much for Coral, who was busy looking after Christine, the twin babies and the dairy. He became restless and got out on the road a couple of times at night. She rang Theo Hill, the only person she trusted enough to look after Harry's horse. Coral gave the black stallion to Theo and Bonnie to use as a sire on their property, Comara Station on the Macleay River headwaters.

Abbey ended his days as a champion campdrafter at the age of nine, in his prime. But what was to follow eclipsed even his reputation as a performer.

In his first season at stud people from all around Australia brought their mares to Comara wanting an Abbey foal. And when the first crop of those foals started competing, they hit like a wave, excelling in events including campdrafting, polocrosse and as top show horses. The Abbey name spread. Abbey became, as the memorial at Willawarrin Rodeo grounds puts it, a legend in his own lifetime.

The Hills moved Comara Stud to another property at Quirindi where the stallion lived contentedly as a sire, loved by the family for his good nature and looks. He died at the age of 27 in 1982.

Astonishingly, Abbey now has more than 40 000 descendants. Joy Poole OAM, who chairs the Australian Stock Horse Society, says he is the most influential horse on the society's stud book.

In early 2013 Abbey descendants accounted for one-fifth of the stud book's 200 000 entries. The nearest horse, Reality, has 17 000 descendants. The list of champions Abbey has produced goes on and on, Joy says. Abbey horses are typically versatile, which has helped them succeed across disciplines. The Willawarrin memorial notes the progeny that triumph consistently in a wide range of equine sports.

Bonnie Hill recalls going to the Warwick Campdraft Gold Cup—the pinnacle of campdrafting—and seeing Abbey descendants take out the first four places. And it wasn't unusual to see them appear prominently among the place-getters at other events, she says.

Joy doesn't think Abbey's legacy will ever be surpassed. 'His line is running as powerfully now as it was three decades ago,' Joy says. 'We hold this horse in massive esteem.'

The society called its premier event the 'Abbey Open Challenge', and the Quirindi rodeo holds the 'Abbey Memorial Campdraft' every year. Willawarrin Rodeo has staged the 'Harry Ball Memorial Open Draft' for 47 years. It's the one everyone wants to win, says Coral. Abbey is also commemorated in the Willawarrin pub in a photo on the wall, and in a tribute song penned by Golden Guitar winner Ian Quinn after visiting the horse's birthplace:

In the hills around Bellbrook and at Five Day Creek your spirit still gallops on by
Born at Willawarrin in the stables at the pub Australia Day 1955 . . .

A portrait of the sleek black horse painted in oils hangs in the Hills' home, and Coral Ball has a painting of Harry and Abbey by well-known artist Rex Newell in the hallway of her home, the first thing she sees as she comes in.

'Harry's dream was to have a great performer—that part of the dream Harry lived to see,' says Joy Poole. 'But Abbey's greatest legacy, that of a sire that has been beyond anything ever seen before or since, unfortunately Harry missed.'

Coral Ball, too, wishes Harry had lived to see Abbey's legacy. The couple had started a list of mares Abbey was to service before he died. That short list seems so humble now in light of Abbey's impact on Australian equine history. 'How I wish that Harry could have known,' Coral says. 'He would have been so proud.'

Written with assistance from Joy Poole, stock horse historian.

18

MADE IN AUSTRALIA: THE WORLD'S TALLEST HORSE

Jane Greenman is a farmer and racehorse trainer. When Jane decided to support an endangered breed of heavy horse, the shire, she had no idea that the adorable foal she bought would break a world record. Jane shares his story.

I've had lots of different breeds of horses and loved them all but I'd always wanted what I call a 'hairy-footed horse'. I looked at Clydesdales but there are a lot of Clydesdales around and they don't need my help. Shire horses, though, are already Category One endangered and there's only 2000 in the world. I thought that if people like me—who don't want to breed them but just want to enjoy one horse—don't buy geldings, the breed will dwindle further. And if younger people don't take over harness as a discipline, a lot of that knowledge about horses and carriages is going to be lost.

A girlfriend in Queensland breeds shires and I thought that was an even better reason to buy one. So I rang Sue and said, 'When you get one of those grey things, those shire horses—it's got to be grey with four white socks—send it down to me.'

She told me there was a four-year wait for them. That would suit me fine, I said, because if I bought another horse my husband would divorce me. He'd already said, 'You've got enough horses'—which was true, I had five. so I couldn't blame him.

Sue rang up two months later and said, 'Jane, it's perfect, your horse is born!' He was born on Christmas Eve—and what a wonderful gift! She'd let me jump the queue because we were both so excited about me wanting a shire.

Six months later, after he was weaned off his mother, he came down to Victoria, a journey that took three days, which is a long trip for a little foal. So this big transporter that normally takes racehorses between states arrives and the driver opens the door. There, sticking out over these panels, were these huge ears, and that's how he got his name 'Noddy' (as in Enid Blyton's Noddy and Big Ears). The driver opened the divider and out came this horse the size of a thoroughbred—and it was hairy!

I said, 'Mate, you've got the wrong horse—I'm supposed to take delivery of a six-month-old foal.'

He said, 'Luv, I believe this is.'

Then I noticed a sign duct-taped around Noddy's neck with big coloured Texta writing, saying, 'I'm only six months old'. Sue had put it on him so he wouldn't get lost because he was the size of a fully grown horse.

When Noddy arrived he was black (all grey foals are born a base black). He would turn a lovely blue-grey later, then start to dapple and eventually turn whiter when he was older. His father, a beautiful grey, is from England—the first shire I fell in love with—and his mother is a jet-black mare.

So Noddy, or Luscombe Nodram, which became his registered name, came to live at our tranquil farm and I was going to learn harness and do some farm work with him to keep him fit and healthy—that was the plan. He was gelded young in life, at eleven months old, because he was going to be a working horse and you don't need a stallion for that. But by the time he was twelve months old Noddy was about 19.1 hands and I thought, 'Gee, he's big, there something's wrong here'.

I rang Sue and said, 'I've got a bit of a problem—this horse is really big.'

'Jane, you know shires are a big breed of horse,' Sue replied calmly.

But I'm used to big horses. I started out in dressage and own warmbloods. And I knew heavy horses don't finish growing until they're about seven or eight years of age. 'No, he's *particularly* big.'

'Look, you'll get over it,' she said.

When Noddy was approaching three years old there was a big horse show in Sydney and Sue suggested taking Noddy because there were only about fifteen entries for shires, there being so few in Australia. He didn't fit in the first truck, though, and we had to find another truck to get him there. I had no idea how to prepare a shire horse for the show ring but people there helped me—plaiting the mane and tail and so on—but we ended up winning the championship! The judge was from England and said that normally young horses couldn't beat the mature ones in the class; Noddy, though, had something about him 'but I don't know what it is', the judge said.

I found out years later what it was: Noddy had another agenda, he was going to become famous!

Noddy broke the records. He was registered with the World Record Academy as the tallest horse in the world, after being specially measured by them. There are strict rules attached, such as that the horse is not allowed to wear shoes at the time. The last time he was measured he was 20.2 hands—he's probably 20.3 now but that's not official. He was equal to the tallest horse in the world when he was about four-and-a-half years old and by the time he was five he'd completely outgrown anybody else.

The amazing thing is that Noddy doesn't look that huge because he's in proportion. If he's standing in the paddock and there's nothing to give perspective next to him, he looks like a reasonably large horse. Then he approaches and his head alone is as big as a human torso.

Everyone jokes that it must be something in the water but, in all seriousness, I did concentrate on his feeding program. I studied vet science nutrition (and genetics) at uni and although I do nutrition programs for other disciplines of horses, these heavy horses are another thing altogether. Because they have cool metabolisms you don't want to 'overcook' their muscles and give yourself problems in the long run. It's very important to have a good nutrition program for them and, of course, exercise. Interestingly, most of the other horses that have been contenders for the world's tallest horse have died at around the age of five because they can't handle the weight on their bone structure and ligaments, but Noddy's healthy and happy, nine years old and still going strong.

He eats three times what a normal horse does. He's got beautiful pasture, so eats as much of that as he wants, and a round bale of hay—equivalent to about fifteen normal bales—will last him a week if we're lucky. Then he has some crushed barley, and vitamins and minerals, now that he's older. His manures are so large I had to get a special pooper-scooper made!

As Noddy grew there were many other considerations and adjustments, too. First of all I had to redesign the stables, getting the welders in to make a special, enormous stable for him. I had a harness made for him and that was going well, and I bought a buggy, but within six months he'd outgrown a full set of harness and the buggy shafts wouldn't fit down his sides anymore, either. I replaced his harness at great expense; the bigger the horse, the greater the thickness of leather and the more strength that's needed. Noddy's gear is three times the strength of a normal harness and was custom-made. The specialist harness maker initially said I'd have to take the horse to him to be measured until I explained it would literally be a case of bringing the mountain to Mohammad.

I had a float made earlier on with double flooring, extra width and length, but Noddy outgrew that when he was one-and-a-half, even though the float can comfortably fit two of my big warmbloods.

Noddy now weighs more than 1.5 tonnes. There's only one truck he can travel in and that's a racehorse transporter, which we call the Noddy-mobile. It has had its floors lowered and reinforced for weight and has been altered so he can move around. It fits three normal racehorses in the space that takes one Noddy.

Everything has to be specially made. He takes an 8-foot long rug, which we pitch over him like a tent. Then there are his halters and lead ropes, all longer with extra thickness and reinforced. I've only had one set of front shoes made by a farrier for Noddy. They cost more than 600 dollars, which is not financially viable, so his feet just get trimmed. Besides, he's not going out on the road so doesn't need shoes. The farrier has kept one shoe and I have the other. Brushing him and shampooing him is a nightmare. I have to set up scaffolding in the stable so that I can reach his back to brush or shampoo it, then pack it away. Yes, he's an expensive horse—thanks, Noddy! Fortunately people love him, because without the help of all those who make this special gear you wouldn't be able to keep a horse like this.

Before Noddy became the world's tallest horse, I showed him to help promote the breed. He was a very good specimen of the shire horse even though he was a gelding. He won all the championships that he could at those shows, not that there are many in Australia. Then he started to get into the category of the world's tallest horse and I had to ask myself, 'What do I do with this animal? And, what was the right thing to do *for* the animal?'

I knew there was demand for him—the horse who previously held the record lived in America and was on the road every day going to supermarkets, shopping centres and arcades. There's big money involved in taking these horses out, and in them advertising breweries or whisky or similar, but Noddy was never born for that and I made the decision early on that he wasn't going to be a commercial horse. I never charge for appearances—I take him to shows for the little kids to love and for people to appreciate horses; for people who have never had the chance to feel how soft and velvety a horse's muzzle is, to experience the beautiful warm smell of a horse and to see how gentle their eyes are.

When we first made an appearance, it was like he was suddenly a Hollywood movie star. I had 150 000 internet hits internationally in ten days. The television stations picked that up and that made it spread. I was on television shows in Australia then on Reuters and Getty Images, which spread the story internationally. I ended up in the *New York Times* and *Chicago Tribune* and the story got into 25 top Russian newspapers. A Tokyo film crew came out to do a documentary on him. His popularity was just phenomenal—Noddy needed a full-time manager to look after him! I was quite surprised in the beginning—although I'm used to it now—because being a farmer and a horse trainer you get on with your job and you're not into publicity and all that sort of thing. Noddy's public relations campaigning has made me happy, though, because it has helped the shire breed not only within Australia but overseas, too.

I pick only the large shows to take him to, because logistically it's hard to get him out. Wherever he goes he has a big pen he stands in with a 2–3-metre barricade around it so that little kids can't crawl through. It's not my requirement—Noddy loves people and can be fenced with a bit of baling twine—but an OH&S requirement of the venues. For him to appear for a week at a show takes incredible effort as I am always by his side. Oh my God, sleeping with him at shows! I've slept with Noddy many times and can he snore! When he gets going, it rumbles. I have to wear double earplugs. He's a typical boy: he loves his sleep, loves his food (licorice is his favourite treat). But he also loves people; he's got a happy outlook.

I love taking Noddy to shows because it's easy to forget how special he is when we're at home on the farm. I get so much pleasure from the little children's faces—they have never seen anything like him. In the five or so years I've been taking him out, and of the thousands of people I've spoken to, there's one man who stands out for me, who really touched me and made it all feel worthwhile.

He was a frail 96-year-old who had come on a four-hour journey from the aged care home he was living in. He introduced himself, saying that he'd worked with heavy horses all his life in a wheat-cropping area, using all the old machinery. He'd read in the local paper that Noddy was appearing in Melbourne and said, 'I just knew before I died that I had to see the world's tallest horse. So I had all the nursing staff and family move heaven and earth to get me here today.' Then he said, 'Dear, I've just come and seen him and I can tell you now I'm going to die a very happy man.' He really meant it. The moment I met him and looked into his face, I thought, 'He's an expert in heavy horses, one of the men who know everything about them. For him to come up, meet Noddy and say, "That is something I've never seen in all my years on this planet," was wonderful. That's when I knew how special Noddy was to other people and that he's really everyone's horse, not just mine.

There was another time that was very special to me, too. I'm never forewarned who's coming to see Noddy and at one show twelve very disabled people appeared in their electric wheelchairs, all with carers behind them. They'd come on a bus together. One of the carers approached me, saying they couldn't get close enough to see Noddy because the crowds were so thick, and that they'd come especially to see him. Could I help? We pushed the barricades aside and all the electric wheelchairs lined up, and I opened Noddy's gate. I realised then that Noddy had never seen an electric wheelchair and wondered how he'd react. Noddy went up to the first person in the wheelchair and gently put his head in their lap and let them touch him, though their poor hands could hardly move. He went to every single one of those twelve people and did that. He was the gentlest giant I have ever seen. I said to my husband later that you can't train a horse to do that, it is in them.

Once Noddy was coming back from Sydney in the transport truck and so was Black Caviar, whom we now call Noddy's girlfriend. It was about the time she was coming up to her seventeenth or eighteenth win, so she was getting up there and doing really well. Noddy, of course, was taking up his three bays so no one could come in beside him, Black Caviar was facing him, so for the whole trip they were looking into each other's eyes, falling in love. She would have been looking *up* into his eyes, though. Black Caviar was dropped off first at Caulfield racetrack. The driver told me that people were waiting around the truck for the doors to open, and everyone was saying, 'Where's this famous horse? We want to see this amazing horse.' So out comes Black Caviar and they say, 'No! We want to see Noddy!' I wish I'd had a picture of them taken together—the fastest and the biggest. Noddy's grey and she's black, it would've been lovely, but I wasn't there at the time. I can always say, though, that Noddy's rubbed shoulders with fame!

I can't remember anyone ever being scared of Noddy on his outings and yet he is the biggest horse they could ever meet. I think it's because he's a light colour, hairy and his name is Noddy; people just come up and hug his legs. I've had people walk over to me and park their baby's pram under him accidentally while they're talking and Noddy doesn't move. It's like he's thinking, 'I don't think I'm supposed to move now.' There's not a nasty bone in his body.

But of course, as with any horse, you have heart-stopping moments. There was one time when someone in the crowd did something wrong. They threw a full can of drink at Noddy, which hit him on the body. He didn't know what was flying out of the sky at him and got very upset. Noddy has been trained with voice commands, but the crowd was noisy and he couldn't hear me. That was one of those times when I fully realise how much weight and height I've got at the end of the lead. Fortunately for me he decided to become an ostrich and bury his head in the sand; unfortunately for me his 'sand' was in my arms. So I held his enormous head in my arms until he settled—it was so heavy that the next day I was in pain from muscle fatigue—but I was happy he got through it. He didn't understand what was happening because he's never been abused in any way, but you can't prepare them on the farm for everything that might happen to them, whether it's people throwing objects or helicopters landing.

But except for this one bad experience Noddy loves going to the shows. It's not that he gets particularly spoilt at them—he's not getting any extra feed—he just loves meeting more people. My other horses go to shows as well but they don't make an effort to meet people the way Noddy always does. Even after eight hours straight and seeing thousands of people he's still happy to keep going—then sleep all night. We've just come from a show in Jindabyne and the Snowy Mountains muster, meeting lots of people from the bush and the city and Noddy wanted to go home with every one of them! He's especially good with kids. Kids don't understand what horses can do and they might have a bit of fairy floss in their hands or something tasty within reach but Noddy would never snatch it.

Noddy likes his work and takes it all in his stride but other than Jindabyne I haven't taken him out for two years because I got tired of it. At the Sydney Royal Easter Show there were 90 000 people a day coming by to see him, which is very intensive for a horse. And me—I admit I'm a quiet, shy sort of person. I still enjoy his appearances but I have to build up my energy for them. When

Noddy does attend an event he needs 24-hour security and has his own security guard as well as me!

I've had offers for him, of course. A horse that famous is going to attract attention because of the potential financial gain. I have had two particularly interesting offers. A country in the Middle East wanted to buy him because they wanted the world's tallest horse and would only need to sell a few more oil barrels to do it! It was tempting in one way to accept but it was highly unlikely that the horse would have lived to the age he's going to live here; he's not intended for life in that climate. So, take the money and run. No, thank you.

The other offer was from America, where a lot of the other tallest horses live. There are considerable financial benefits to owning such a horse in America, but I don't think it's a life for Noddy—he'd have to make appearances like a movie star just about every day and a horse has to live as a normal horse. Here, he can breathe fresh air and relax. Noddy was never made to make money; he was made to bring people pleasure. And I do it for the love of it. I take the responsibility of owning the world's tallest horse very seriously but in fact I'm like a caretaker—he's Australia's horse. We made him, and those who've met him love him. I take it as a compliment that the offers were made but I couldn't do that to Noddy.

At home, he's just one of the family; being the world's tallest horse doesn't mean the other horses pay him any special attention. Age seems to be more important in the paddock. The warmbloods are smaller than him—every horse is!—but they're the boss males on the farm, and my little quarter-horse mare who's only 15.2 hands runs rings around him. Noddy well and truly knows the hierarchy. The only horse younger than Noddy is a Suffolk Punch, a really endangered breed, and he's the lowest in the pecking order.

Noddy was trained to harness but we don't do carriage work because he doesn't fit any of the shafts and the wheels are always too small; no matter how big the carriage is he makes it look like a little sulky.

I did, however, learn how to long-rein him, which was hilarious. Long-reining is when you drive the horse walking behind it using extended reins. The trainer told me to keep an eye on his head, to watch his ears and his eyes to make sure he was concentrating, but all I could see was this huge back end—muscled, lovely and well groomed! So everything I do with Noddy has to be pure voice command. I've never seen what's going on up front but Noddy's beautiful, he knows what he's doing. I long-rein to work him, which is good fun and I get exercise at the same time.

We do ride him, too, though I used to have a ladder to get on him and my normal saddle looks like a jockey's saddle. When I finally get on I joke that I'll get altitude sickness because when I look down the ground's a long way away. When you ride Noddy it's like steering the *Titanic*—there's so much horse underneath you that turning takes much longer and you need more area to do it. Then when I dismount it feels like it takes ages to reach the ground.

When Noddy canters you feel the ground rumble; it's like thunder, or being at the races when the horses gallop past the finishing line. To ride him is lovely, though. He has a very smooth gait but it feels slow because the strides are so big. The shires aren't really designed to keep up with other horses; they're all about slow, steady strength and power. When farmers used to ride heavy horses in the old days it was only to ride back from the paddock to the farm after a day's work. You wouldn't want to ride a horse that big in a trot or canter often because you could damage its legs. Noddy's really a walking horse. I'm always mindful of that. Occasionally he trots, though never on hard ground, but that's not how he gets his exercise. And I would hate for anything to happen to him.

You never want a horse like Noddy to fall. A horse that big can't be operated on—you can't anaesthetise him because of his body mass, and cardiac arrest is also a problem under anaesthetic—so I want to avoid accidents at any cost. You can't ever have the horse gaining excess fat, either. Where a normal horse puts on excess weight it might be 30 kilos; with Noddy you wouldn't even notice an extra 200 kilos, but that's a lot of weight to carry around on those joints.

I took him trail riding with friends a couple of times, thinking I'd give him some exercise out in the bush. We were going pretty well on the first ride until we came to an entry area to the park, which is designed to stop cars but allow horses in. It has these barricades you have to ride between like a little zigzag but Noddy and I got stuck because he was so broad we didn't fit between the posts. Fortunately, he's been trained so that whenever things don't go according to plan he stands still until someone gets him out of the predicament. So Noddy waited till I lifted up one foot and got it over the rail, then the other foot, and we were able to get out of the situation without any damage. On another trail ride my friends were ahead on their horses and next thing I knew I didn't have a horse under me—I was hanging off a branch up a gumtree! Noddy was fine, he just kept walking, while I yelled, 'Excuse me, could you bring my horse back!' My friends led Noddy back under the tree and I dropped down onto him. So while I've never technically fallen off him, I guess have come off him. Anyway, that was the end of our trail riding.

He has accidentally trodden on my foot once and it wasn't his fault, it was mine. Because his legs are so long, when he walks you have to give yourself extra room in case he steps sideways into your line of walking. He was just walking along and he slipped on a stone with one foot and put the other foot down to get some balance. I wasn't concentrating and suddenly there was the world's tallest horse on top of my foot. He didn't even realise, thinking it was just part of the ground. I gave him the command to move off but I knew I'd broken my foot. I remembered the rule that you should never take your boot off until you get to the doctor because of the swelling, but when I got home I realised I couldn't drive so I pulled off my boot, duct-taped my foot up, took a few painkillers, went to bed and never actually went to the doctor.

It's times like that when I'm reminded that Noddy is the world's biggest horse but it mostly doesn't cross my mind because I'm just his 'mother' and he's just the same as the foal with the big ears who got off the truck that day.

19

THE HORSE THAT MADE THE MAN

In 2008, pacing trainer Gary Hall answered a call from an agent whose name he didn't recognise. The agent was phoning from New Zealand and wanted to know whether Gary would be interested in buying a horse by the name of Themightyquinn. Gary, one of Australia's leading trainers, hadn't heard of the horse but did his homework, watching clips of the gelding in his most recent races. The horse had notched five wins from 26 starts, and a third in New Zealand's premier race for three-year-olds. Not bad.

Gary found out that the gelding was being passed over by other buyers at the time; at 14.2 hands high and slight of stature, he was small for a pacer. Dark with no white markings, he was, to anyone else looking at him, just a plain little horse. But Gary liked his big stride and the way he pulled; his stable was good at working with horses that pulled. He paid 180 000 dollars for the gelding—more than he'd ever spent on a horse—buying him on behalf of seven owners. The group imported the pacer and renamed him Im Themightyquinn to avoid confusion with an Australian horse that had a similar name.

Gary had trouble with his new charge at first. Couldn't control his urge to go flat out from the barrier. He used a rubber bit, which helped 'Quinny' stop fighting the bit and to settle. Once he was persuaded to relax, the horse proved a joy to train. Gary trained him solidly, working him in sand at his property at Hazelmere in Perth, building up his aerobic capacity, the way he works all his pacers. (Pacers, unlike trotters, move in a lateral gait, with the front and hind legs on the same side moving together.) Gary says his key to success is simple—you just concentrate on the fitness and wellbeing of the horse. He's not big on devices like heart monitors or the latest food supplement—how a horse looks in the eye tells him how well it is. He's patient, too, doesn't push a horse if it's coming back from injury or illness and gives it longer than it needs to recover.

As Gary spent more time with him he came to appreciate Quinny's character and foibles. He's a trusting horse who'll do anything you ask of him, and do it kindly, the trainer says. Smart, too. He's a gentleman on the ground, never pushing anyone around, and maintains his distance, only ever coming over to you in a yard or paddock if he wants something, such as a carrot from Gary's wife Karen, one of his owners. He's regimented in everything he does, sleeping at the same time every day and starting the day with a morning routine that drives Gary mad: once he's harnessed he'll walk ten or twelve paces, stop and survey the scene, pass manure, then walk off again slowly, and will only work if he sees another horse training. As he started racing him Gary Hall realised something else about the pint-sized gelding: he had an absolute will to win.

Quinny won his first race, the McInerney Ford Classic, at Gloucester Park in Perth in late 2008. But he came third in his next start, which disappointed Gary and taught him a lesson. The gelding had led the race by 4 lengths then been overtaken; Gary realised he performs better when he comes from behind. His first big win was the Fremantle Cup in late 2009, then came the Cranbourne Cup in Victoria a year later, a triumph that put Gary Hall on the map Australia-wide. Within three months of that race, in January

2011, Quinny won the Fremantle Cup again and two weeks later the WA Pacing Cup, then the Auckland Cup in March that year. It wasn't long before he was being talked about as the fastest pacer in Australasia.

Some horses need to be persuaded to win but Quinny always wants to get in front, says Gary, whether he's 2 lengths behind the leader or 5 lengths. He'll emerge from the back of a bunched field, go out three or four wide and blitz the others in an explosion of speed. He powers around them looking like a much bigger horse. Like he's got wings on his feet.

He's driven most of the time by Gary's son, Gary Hall junior, a top reinsman. The two go hand in glove, says Gary senior. The horse will win for others but seems to do his best with his son. A brilliant driver, his father says. Gary senior used to drive the 'bike', as sulkies are called, himself, winning more than 300 metropolitan races. Thought he was pretty good till Gary junior came along and he realised that he was better as a trainer. His other son Clinton followed in his footsteps, too, and is now Gary's foreman at the stables.

Im Themightyquinn's performance in the 2012 Fremantle Cup was nothing short of spectacular. He was running second last as the bell for the last lap sounded. As the crowd watched, waiting for his trademark late sprint, Gary junior edged him out five wide then bolted past the others at a rate that had to be seen to be believed. The crowd went wild. It was the third year in a row that he'd won the race.

But the Interdominion Cup is the big one in harness racing. Run since 1936, it rotates around the Australian states, and the north and south islands of New Zealand, building up to the main event with trials in the fortnight beforehand. The usual distance is more than 2400 metres. Quinny won the Interdominion in Auckland in April 2011, Gloucester Park in March 2012 and in Sydney in 2013, equalling the record for winning it three consecutive times.

The 2012 race was the most thrilling of all for his trainer. Quinny was running last. Gary waited and waited for his son to make his move, convinced for most of the race that he was never going to do it. Gary junior edged the pacer out wider and wider. The field turned into the back straight. There was no way he could get up from there, thought Gary, wondering what on earth his son was doing and already thinking of how he'd tell him off. Then, a mere 600 metres from home, he made his run. Gary Hall watched in shock as the Quinn surged past the other horses, led the field emphatically and flew past the post. By then the trainer was whooping and yelling along with the rest of the hometown crowd. 'The Mighty Quinn! The Mighty Quinn! The Mighty Quinn!'

The crowd clapped horse and driver off the track. Gary had never seen a tribute like that before. He was overcome.

People love Quinny, though sometimes they're surprised to see him in the flesh, expecting a more impressive looking animal, he says, laughing. Gary is now recognised in public because of the horse. He and Karen pulled up at a roadworks detour in Perth recently and wound down the window to find out from one of the workmen where they needed to go. 'How's the Mighty Quinn going?' asked the worker. On another occasion Gary wanted to buy a ride-on mower and the man in the shop asked him if he could deliver it—and have a look at Quinny at the same time. A set of shoes the horse wore in one of his wins sold for 5000 dollars in a charity auction.

There have been disappointments and scares along the way, though.

At a time when the Quinn was winning races and doing his Perth trainer proud in the west, he ran last in the Victoria Cup in December 2011. Gary found out later that he was dehydrated and couldn't perform. Quinny came down with a fever the night before the Interdominion Cup in 2012. Gary was beside himself and couldn't sleep and kept checking the horse through the night.

Apart from anything else the Cup was worth 600 000 dollars. But Quinny rallied overnight and won it. He's a tough little bloke, says Gary.

Three months later, following a spell after the Interdominion, he was competing in some trials when he stopped short after 60 metres. Gary watched from the side of the track, alarmed as the horse staggered a few more steps and couldn't go on. He froze; he'd seen horses drop dead after a heart attack or when they'd 'bled out' (burst a blood vessel). He and Gary junior walked the horse off the course as quietly as they could and put a stethoscope on him. His heart was racing wildly—he'd had atrial fibrillation. The harness-racing world was reported as being 'stunned' at the breakdown and there was talk of doubt surrounding his racing career. But Quinny recovered well after Gary rested him.

In October 2012, Im Themightyquinn was named Australian Horse of the Year by Harness Racing Australia. By the end of that year, he'd won 3.4 million dollars, eighteen consecutive races and was undefeated over three years at Gloucester Park. He'd notched 44 wins and 27 placings from 89 starts. Within three months of that he'd pushed it to earnings of 4.1 million dollars and at the time of writing was being talked of as Australia's richest pacer. Gary says he doesn't expect to ever find a better horse.

Life is sweet for the irrepressible Gary Hall. Gary, aged 63, has 44 horses in work and has just bought a hundred-acre property in Hazelmere to expand his stables. His first property consisted of 5 acres of dry, reclaimed sheep paddock with a sandy track round its perimeter to work the horses; the new one has green paddocks, a 600-metre sand track, 1200-metre jogging track and a 1000-metre oval fast track similar to those used in races. Gary's often pictured in sporting pages smartly dressed in suit and tie with the owners of the night's winner gathered round a sulky at Gloucester Park, beaming, arm around Gary junior. Quinny is only one of his winning horses. On one night Gary equalled a record set at Gloucester Park, winning six out of ten races.

He's proud of all four children. He's happily married to Karen, his fourth wife. The couple owns shares in twenty horses. Gary Hall, who as a young man was neither sporty, academic nor ambitious, was Perth's leading trainer for seven of the eight years to 2013. He was proud to be nominated for the Western Australia Hall of Fame in the racing industry in 2012 and hopes to be in it one day. Life doesn't get much better.

But it's been a hard climb to the top and despite everything he's done, it wasn't Im Themightyquinn that got the trainer there. Gary says he owes it all to another horse: Maru Adios. He's the horse that holds his heart.

Gary Hall's love of horses in harness goes back a long way. He first drove a horse in long reins when he was eight. He used to get out of bed early in the morning to go out to the kerb to pat the local milk horse, a feathery-legged animal with his head in a nosebag. After a while the milkman let him drive the cart. Gary was a natural. Either that or the milk horse drove itself, he says with a smile. He was fascinated, and loved the look and smell of the horse.

He'd got the horse bug as a child of five reading *Black Beauty*. He'd sit on his grandmother's bed as a young boy listening to her talk about life on the family farm in the wheat belt northeast of Perth, ploughing the team of horses in the fields or riding down to the store in Merredin. He first rode a horse at a friend's farm at the age of thirteen, assuring the friend's father that he had ridden before. The horse took off and careened to the other end of the paddock. It was a thrill. He went riding on weekends for a couple of years at a riding school called Tappers, near North Beach, where you'd pick out a horse from a yard of old trotters and retired racehorses and ride out in the sand dunes and bush.

Gary first went to the 'Trots', as they are called, as a fifteen-year-old in 1964. He'd already been betting on the races with a school friend, Frederico, who introduced him to gambling and, looking a bit older, would place bets for them both at the TAB. The pair would then listen to the races on the wireless. The commentator's calls used to deteriorate according to how much he'd had to drink before them, Gary recalls. Frederico suggested going to Gloucester Park one Friday night to watch a race. Gloucester Park was only a ten-minute bike ride away and as Gary slept out on the verandah of his home in summer it would be easy to steal away for a race or two then get back into bed. He left home in his dressing gown. The pair, being underage, had to sneak in, jumping over the fence to the track after the guard patrolling its perimeter had walked past. In those days, about 30 000 people went to the Trots and Gary would sit in the wooden grandstand among them, exhilarated. Frederico placed the bets for him from the money Gary earned each week selling newspapers or fish he'd caught or from refunds for bottles he collected at the football.

Gary decided one night in 1965 to put all his winnings to date—20 pounds—on an outsider. He'd watched the horse race before. The odds were 66 to 1. It won. He took home 1320 pounds—a huge amount of money in those days, more than what his parents' house was worth. He rode home on his bike with the cash tucked beneath his dressing gown held in place by the cord. Back at home he was counting the money on his bed when his mother, who'd discovered he was out and heard him return, investigated. She burst into tears at the sight of what could only be ill-gotten cash and called out to his father to come and have a word to their son. His parents were relieved when Gary finally admitted where the money came from. It wasn't long after that that they started going to the Trots with him and soon became avid followers. He can't remember what happened to the winnings—he says he would have handed it over to his mother and maybe got a new bike—but in general gamblers don't do anything with their money other than gamble with it, he says, laughing. It's the same with him with horses now—if he wins with them he buys another horse.

An idea was emerging as he watched the trotters fly around the Gloucester Park track as a teenager: 'I can train horses.' He was inspired by what he saw: the gleaming horses streaking round the track with the crowd cheering them on. The 60s through to the 80s were a golden era for trotting, when crowd attendances were at their highest. Gary loved the horses, too, and wanted to work in an industry that involved them. He was still at high school, however, and his father insisted that he complete his Leaving Certificate. He did, and joined a stock firm when he left school and trained as an auctioneer—the start of a string of jobs in which he was completely disinterested but which funded his dream of becoming a professional trainer.

It was while he was working for a farmer who raced standardbreds that he bought his first horse, an aged trotter called Mukaboy. He started at reinsman school with Mukaboy, buying him on a Saturday and showing up for his first day the following Tuesday. The school was having trials that day—a prospect that terrified the new student. He realised that he had no idea of what he was doing. The trainees lined up in their sulkies for a standing start and Gary vaguely recalls that Mukaboy 'broke'—or galloped—early, ruining their chances.

He got his licence after twelve months at driving school and started training horses he leased, getting up at 3.30 in the morning to do it before going to his day job. He moved from job to job—hardware store, feed store, a bank, the public service. He used to help out with the trotters at Tramby Lodge on the Swan River in his spare time, where they'd jog the horses on the old aerodrome nearby. In return he was allowed to drive in some races at Northam, a regional track. His first race there, in early 1968, was driving a horse called Tiw. He came fifth and was pleased. He had his first win with Russvic in October that year, leading all the way.

In 1971 he notched his first city win at Gloucester Park with a horse called Tobaree. His successes training other people's horses built up. In 1978, Gary and his third wife Sue bought the 5-acre property at Hazelmere and set up a training establishment. The couple had three young children by then: Clinton, Gary and Tenille. Gary leased out the property at first, scored a few more successes, and announced to Sue that he wanted to leave his job and train horses full-time.

'I didn't realise how hard it was.' The pressure to break into the industry and support a wife and three children proved enormous.

Gary worked full-time for an owner, a lawyer who had a few horses at his property at Sawyer Valley, where he moved with his family. But he couldn't make a success of it. The quality of the horses wasn't up to scratch and the property with its rocky, hard ground wasn't conducive to training. As a trainer you pay for the feed and upkeep of the horse and split the profits from the winnings fifty-fifty so if the horses don't get placed you're out of pocket. As their situation became dire, Sue announced that she was moving back to their house at Hazelmere, and Gary agreed that they would all go. They returned to Hazelmere in 1983, broke, and with Sue pregnant with their fourth child, Casey. That week Gary sold his mother's sewing machine for 38 dollars to pay for food.

He had been unemployed and on the dole for three months when Maru Adios appeared. A friend of Gary's had been leasing the horse and wasn't doing well with him. Gary asked if he could try training him and the friend agreed, handing over the lease. That was the turning point in Gary Hall's life.

At the time, no one thought Maru Adios was any good. He was eight years old and out of form, a cast-off that hadn't notched a placing for three years. He looked like an 'upside down, pregnant pasty', Gary recalls—flat back on top with a round belly below. A bright bay, he had a likeable face though with a Roman nose. Gary clicked with the horse right away. He set about returning him to form.

A self-taught trainer, Gary was developing his own, often unorthodox, methods. He used to ride Maru in a saddle through heavy sand. It showed him a way of training that others weren't using and it was effective. He worked the gelding for months, until he was fit and well. 'We ran his little legs off!'

Not long after he got him, Gary 'oiled' Maru to get rid of any sand in his stomach. He mixed in a beer bottle full of linseed oil with some paraffin. He should have used a cupful. 'I nearly killed him.' The horse was sick for four days, couldn't hold his food. The vet said there was nothing they could do for him but wait. Gary was stricken. He lay down next to the horse at times to make sure that he got through it. And Maru came good. In fact, he came up better than ever. Cleansed him out. His coat shone and he was dappled from head to toe.

Maru Adios started winning races within five or six months. A chance meeting with an old man in Kalgoorlie helped. One night in September 1983 Gary and Sue were so desperate for money that they put the four children in the car and Maru in a float and drove seven-and-a-half hours to the race meeting in Kalgoorlie. The race offered a first prize of 2800 dollars and a second place of half that. The way the system worked Gary would have to split the first prize with his co-owner, but could collect the entire second prize as trainer; either way he'd take home 1400 dollars. Maru Adios ran second. Gary was happy and relieved. As he went to unharness the sweaty horse, a man approached him, drunk. He was about 60, bleary eyed and unkempt, and was slurring his words.

'This horse is never gonna win unless you put him in a hood and earplugs,' the man said.

Gary looked at him, sceptically. 'Yeah, and what would you know about it, mate?' he said.

'I know because I used to own this horse in South Australia,' the man said. 'You try it. Mark my words.'

Gary took the advice the next time he raced the horse. It couldn't hurt, he figured. The race was at Gloucester Park and he'd drawn barrier nine—the widest possible barrier. He had virtually no chance. He put on the hood and deafeners, which block out the surrounding noise and calm a horse. Maru Adios won. He's used the earplugs on his horses ever since. 'You can learn from anyone,' says Gary.

Maru Adios kept winning and getting placed. His earnings began paying the household bills. In twelve months or so, he won eleven races and scored 22 placings, netting 38 000 dollars. Gary was back in the industry and on his feet financially again, though not enough to be able to keep Hazelmere, which he and Sue were forced to sell and then rent back.

The pacer started to be noticed. His owner ran him in many 'claiming races'—races where people have the option to buy the horse. One fateful night in late 1984 after Gary had been training him for about two years, he ran second in a race at Gloucester Park and was unexpectedly bought. A steward came up to Gary after the race and told him that Maru Adios had been claimed. He was put in a float and driven off. Gary was devastated as he went home without him. It was like losing a member of the family. 'I loved that horse,' he says.

Maru Adios never won another race. He broke down not long afterwards with an injured ligament, meaning it was unlikely he'd ever race again. The new owner rang up Gary to say that he was going to have the horse put down and asked if Gary wanted him. Gary had other horses that were doing well for him by this stage but couldn't afford to support him. A young woman who helped out at the stables said she'd take him as a hack. 'I'll look after him,' she promised. Gary arranged for the horse to go to her on the condition that she looked after him. He went to check on Maru Adios six months later and found him thin, lame and in pain with a condition called 'greasy heel', a herpes-like disease that was making its way up the horse's leg. Gary arranged to have him euthanased. It was another blow.

Gary kept improving as a trainer, going on to work and co-own Zakara, which won almost 500 000 dollars. He also owned a share in the New Zealand pacer The Falcon Strike, which earned more than a million dollars before he was retired in 2006. Then came Im Themightyquinn. In and among them have been other successful horses.

But, for Gary, none has been like Maru Adios. He says he wouldn't have gone on to be the trainer he is without him. In fact, he wouldn't even be in the industry if not for him. It's not just what he did for Gary and all the successes since, Gary says, 'He was just a beautiful horse.'

PART VI

Horses that help

20

SALLY'S HEALING HORSES

As a girl growing up on her family's riding establishment, Tooradin Estate, Sally Francis often witnessed impressive displays of horsemanship. Her mother, Judy, was a superb equestrienne, twice runner-up in the Garryowen, Australia's most prestigious show-riding event for women held annually at the Royal Melbourne Show, and a fine all-rounder. Her father, Derry, was a stockman and an A-grade polo player. Her parents hosted competitions including cross-country events, attracting other top riders to Tooradin. There was much to inspire her.

Sally remembers watching one of the pupils at the family's riding school, a young boy with cerebral palsy, ride around the arena, doubled by an able-bodied rider. The boy, with a group from the disability service organisation Yooralla, was usually in a wheelchair and she'd seen him walk awkwardly whenever he was out of it. Yet Sally was struck at the way he moved so easily on the horse and how elated he was doing it. She was taken, too, by the way the Yooralla group revelled in the freedom of being outdoors, laughing even as they were caught in a sudden downpour; for some of them probably for the first time in their lives.

Several decades later Sally runs Tooradin Estate and teaches classes of people from Yooralla herself. A Riding for the Disabled Association (RDA) Victorian coach for 25 years until 2011, Sally has made it her life's work to help other riders, including assisting riders with disabilities from the grassroots to the elite level of competition, here and overseas. Last year, as team manager of the Australian Paralympic Equestrian team at the London Paralympics, she watched and cheered along with the rest of the team as Joann Formosa took gold. As impressed as she is by the abilities of the riders, though, Sally also knows that it takes a special sort of horse to get them there.

Sally has twelve horses that she entrusts with the care of riders in her Yooralla classes and never ceases to be amazed at the way they behave around them. The horses need to be unerringly steady as even the slightest, most innocuous movement can put a nervous rider off—a pony that shakes its head or snatches the reins, or the sight of another horse approaching with its ears back. Sneezing, stumbling or neighing can all frighten the riders, whose wellbeing can depend on having a very stable environment.

The sessions can prove a stressful environment for the horses, too. The arena might be crowded and noisy and filled with the people—people leading the horses, helpers walking alongside the riders, instructors and physiotherapists, and riders who scream, cry or shriek with happiness. It can be demanding on a horse having a rider who's not well-balanced, whose body spasms involuntarily or who gives confusing signals.

Sally makes sure, firstly, that the horses are suited temperamentally to the task. Many of them are aged—she jokes that Tooradin Estate has become a retirement home for horses—but it's more a matter of character. She trains them, making sure they stand patiently at the mounting ramp designed to take people in wheelchairs or those who can't walk well up to the saddle. She checks that the horses can adjust their walk when required, have a slow jogging trot, are easy to ride and work well in a group. Then she watches to see if they're happy doing it.

The same horses are used in her other riding lessons but Sally has noticed they're better behaved for the Yooralla riders. Take Mossy, the quarter horse–Arab cross, who can be tricky for able-bodied people to ride as he's sometimes nervy and shies. But when David, who has Down's Syndrome, rides him, he'll calmly go anywhere he points him.

Royce is a six-year-old purebred Highland pony not usually ridden in Yooralla classes as he can be nervous. But Royce loves to be brushed, so when a young woman with autistic spectrum disorder came but was too nervous to ride, one of the coaches gave her Royce to brush. An hour later the woman came over to Sally, who hadn't yet met her, and touched her on the shoulder. She pointed to Royce then walked Sally over to the mounting ramp. She seemed so keen that Sally helped her on. Royce stood like a statue. It was so important that he did that for this particular rider, she says. And when he walked, he slowed down his normal pace.

When the class finished, Sally was talking with staff from the Yooralla centre who were saying how pleased and surprised they were that the young woman, and another similarly anxious girl, had both ridden Royce. The carers hadn't expected them to overcome their nerves to be able to ride, especially on the first session, but Royce had won them over. As they talked, the young woman, a loner who didn't communicate well, came over and stood beside Sally and put her hand on her shoulder. The staff were even more surprised at that.

Beau, the Welsh mountain pony with the beautiful big eyes, had a similar effect on Craig. When Craig first arrived at Tooradin he was reluctant to even get out of the centre's bus. By the end of the session with little Beau—a riding school favourite—he was smiling and giggling in a way that the Yooralla staff had never seen; Craig is usually non-responsive.

Sally loves and appreciates all her horses, but one stands out: Strauss.

Sally and her aunt purchased the pretty steel-grey Anglo Arab with the look-at-me presence and 'magical' trot for Sally to compete on at shows, showjumping and, her passion, eventing. (Sally was a keen competitor and once part of a national team in a friendly eventing tournament in Germany in the early 1980s.) Sally's aunt would go to the competitions with her, a wonderful time for both women as her aunt was elderly and nearing the end of her life.

Fours year after she got him Sally enlisted Strauss to help her obtain her EA (Equestrian Australia) Level One coaching certificate, which then enabled her to get qualifications for RDA Victoria and RDA Australia. She took him to national RDA workshops for riders and carried out lunging and jumping demonstrations with him. During one of these workshops she offered to loan her horse to para-equestrian Sue Harris, who was also heavily involved in RDA as a state coach. Strauss helped Sue qualify for the Sydney 2000 Paralympics team.

Sally also loaned him to a rider with cerebral palsy in Queensland for the para-equestrian national championships when the team in that state was having trouble finding a horse for her. The rider had poor torso control and balance and fell off Strauss a few times at first; his natural trot was too forward for her. He slowed it down. Sally recollects watching them compete in a dressage test at the national championships. The rider's hands were very unsteady and she was showing considerable involuntary movement as they were coming down the long side of the arena towards Sally, yet Strauss was trotting perfectly, keeping up an 'amazing' rhythm and tempo. Then, as he approached, he looked at Sally as if he were saying, 'I'm doing the right thing, aren't I?'

Sally cried.

'These horses seem to be incredibly understanding and want to help people—that's what's so special.'

As an RDA state coach and national coach assessor for RDA Australia, Sally came across some remarkable horses elsewhere, too. An Australian stock horse at one riding centre could canter beside an instructor as they walked, even off the lead, making smooth transitions as he did. He was a great horse for young men who'd become paraplegics or quadriplegics, who didn't want to take part in games with the others, and who could ride at a canter in a stock or Western saddle.

Tesoro was a different case altogether. Sally was in Singapore for three months helping train instructors for a new RDA centre when she came across him. The Singaporean instructors would warm up the horses before a session then line them up ready for the riders, calling them forward one by one for each rider. They'd call Tesoro forward but if he didn't like the look of the rider he'd prance and pigroot, and they'd have to lead him away and put him back in the stable. But if Tesoro liked the rider, he'd go and stand next to them. 'He'd worked out who bounced around on his back!' says Sally. At the end of the session the instructors would always line up the horses again, but some of the mares got wise to this and would work for about ten minutes then go and line themselves up.

Sally, who has also been a long-term pony club instructor, says she always assesses people quickly, regardless of who they are, then concentrates on their abilities. Known for her generosity in sharing knowledge and innovative ways, she always thinks 'outside the square' with the people with disabilities who come to ride at Tooradin to help them develop the best use of their reins and legs, and the best communication with the horse they're on. This might mean using aids: rubber bands that secure loose feet to stirrups yet break if anything goes wrong; reins with rubber tubing to identify where they should be held; or a back protector to help support someone with poor torso control.

Sometimes it's a matter of trying different approaches. Many participants are keen to ride but scared of the height from the horse or the ramp to the ground. A ten-year-old boy with Down Syndrome who got to the top of the ramp recently became terrified, refusing to look at the teachers and crying, 'No, no, no'. Sally turned it into a game, running round the ramp and getting on and off the horse until the boy could see that it was safe to do this and became used to the idea. He eventually was helped aboard, and beamed.

Sally has watched many times as the fear melts away when a horse first moves and the new rider has a look on their face that says, 'Wow!' She's seen the joy and amazement, too, as someone moves free of their wheelchair for the first time in their life or sits tall in the saddle rather than in a chair looking up at people, then as they feel their muscles unwind as their body rocks gently with the horse.

There's something about the warmth and motion of a horse that's comforting to anyone who likes horses, but there are physiological reasons why riding helps people with disabilities. The movement of the horses' hind legs helps people who have difficulty in mobility to 'pattern' their body to move, improving muscle tone, muscle strength and coordination when they walk afterwards. Riding helps posture, is good for mobilising the pelvis and for shifting weight from one leg to another. 'It's a physiotherapist's dream treatment!' Sally says.

Riding benefits people who are 'non-verbal', too. Sally has had riders whom she's never known to speak before come out with a word or command for a horse. It might be someone who loves trotting who learns that they can get the horse moving faster by saying 'Go'. One rider with acquired brain injury, who was learning to speak again, struggled for months then finally got out the command to get her horse to canter. 'Fucking canter,' she said, with a big smile, and the horse obeyed. What about when she wanted to trot? 'She only ever wanted to canter,' Sally explains, smiling. 'She'd go from a walk to a canter.'

It doesn't always work out, though. Sally says that one young man with autistic spectrum disorder, a big chap, had been so keen to ride that he'd lost 30 kilograms so that he could do it. He arrived one day, gung-ho and ready for action. Sally had a wonderful ex-eventer called Ned ready for him at the mounting ramp, but when he got halfway on the horse he decided he didn't want to be there and panicked. He tried to get back onto the mounting ramp, pushing with all his might against the saddle with his leg. Ned stood rock solid, although the weight of that push should have shifted him sideways. 'We gave Ned lots of strokes and pats to thank him after that,' says Sally.

The would-be rider hasn't been back but Sally says that when he does return she'll try another strategy to help him feel safe. If someone really doesn't want to ride, she might suggest learning to drive a carriage, doing a grooming program or groundwork with a horse at liberty.

In the time since she was a girl, opportunities for riders with disabilities have changed dramatically in the equine and wider community. Pony clubs and adult riding clubs have become more inclusive and Equestrian Australia has incorporated events for differently abled riders. Sally goes to pony club workshops advising on 'inclusive coaching' and techniques for teaching people with disabilities, and is on the RDA board. In addition, she oversees the activities of three riding clubs at Tooradin and events including cross-country competition, 'equine touch' workshops, 'instructional' trail rides and Equine Facilitated Learning groundwork on the 400-acre estate.

Her work assisting riders with disabilities at the elite level has taken her around the world. She has been involved with the Australian para-equestrian teams since 1994, travelling with them to the World Championships in Hartpury, England, as a carer; then as team manager or 'chef d'equipe' at the Atlanta Paralympics; assistant coach for the team at Hong Kong in 2008; and team manager again at the London Paralympics. Yet Sally still marvels at the sight of someone in a wheelchair bumping across a paddock to greet one of her horses with a bridle, and more so as the horse lowers its neck, carefully and slowly, to put its head in it.

21

TO THE RESCUE

It was 5 June 1979. Tasmanian Terry Fenton (see Chapter 5) had been riding for more than five hours, scanning the bush for a small boy in a yellow jumper. It was almost midday, and he and another search-and-rescue rider had been walking their horses through the scrub since dawn looking for the missing six-year-old.

Clint Bolton had disappeared from near his family's shack at Beechford on Tasmania's north coast nineteen hours before and there were still no signs of him. The boy had been running ahead of his parents as the family was coming back from the beach in the afternoon when he became separated from them. His father had watched as his snowy-headed son had appeared and disappeared in the sand dunes in front of them. Then vanished.

A massive search started that evening, drawing more than a hundred people: police, locals, volunteers from further afield, the State Emergency Service (SES) and an aircraft. Nine members of the Gee Park Mounted Search and Rescue Group saddled up after dawn the next day. It was their first official search as a team.

Terry, president of the group, rode with mate Terry Davis. They'd been assigned a grid on the map near the shore, moving between scrub and open flat coastal country. As they rode Terry wondered about the boy. He was in his early 30s and had kids himself. It was winter and the night before had been cold, dropping to around 5 degrees at the coast and near zero inland. The boy could have hypothermia at that temperature, could have stumbled and twisted an ankle, slipped in a creek or fallen off rocks. The searchers at the operations base at a recreation reserve at Beechford speculated, too, about what had happened to the primary-schooler; how he'd got out of earshot of his parents so quickly, whether he was injured and about his chances of surviving in the wintry weather.

The two horseman pushed on. Terry was hopeful. He'd been out on a similar search previously, which had turned out well. That was in the summer of 1973 when another, younger boy had gone missing after wandering away from where he'd been playing in Prospect, then a bushy suburb in Launceston. Terry had been approached that afternoon by policeman Bill 'Bluey' Bessell, who ran the one-man station at Prospect, about joining the search.

'Haven't got a spare horse, have you?' Bill had asked. 'Could you lend us a hand? A young fella's been lost.'

Terry was president of the Gee Park Rodeo club at the time and asked another member to join in, searching into the night with him and Bill, to whom he'd loaned a horse. They returned early the next morning with more members of the rodeo club to resume the search. Some shooters found the boy just as Terry and another rider were approaching on horseback. After the success of that operation Bill suggested forming a permanent team. The Gee Park Mounted Search and Rescue was established, with Bill coordinating their part in any searches.

By the time the search for Clint Bolton took place the squad had a core of eight volunteers, ever ready to pack their horses and gear into floats and head out to where they were needed.

In Clint's case they made all the difference.

At 11.45 a.m. Terry Fenton saw the small boy in the yellow jumper.

'Hey, help, I'm lost,' he said. 'You're not looking for me, are you?'

'Is your name Clint?' Terry asked.

The little boy's face lit up.

Terry lifted him onto his saddle and rode back with Terry Davis to a clearing where the boy was put in a small plane and taken back to base. Clint was excited about the horse-ride, but more excited about the trip to Beechford in the plane, Terry says, laughing. The two riders, who had found him 10 kilometres from where the family had been walking, joined the others in the Gee Park squad and cantered as a group into the operations base to a rousing cheer. Terry, an unassuming sort of bloke, remembers the lunch put on by the Salvos more than the accolades.

The mounted squad proved ideal for such searches. The riders could cover four to five times more ground than people on foot, go further, higher and for longer than walkers, and carry more. 'There weren't too many places we couldn't take the horses.' The riders were higher up than walkers so could see better, too. They were all seasoned bush riders, well used to Tasmania's rugged conditions and treacherous terrain. Conditions in the central highlands and lake area in particular could turn hostile. The weather could be notoriously unpredictable, closing in and deteriorating rapidly. It could be summer and T-shirt weather at one o'clock then snowing by four, says Terry. The ground in the highland country is stony, with drains and ditches that might only be several inches wide but 3 feet deep. Other areas could be packed with bogs. It took the team nearly three hours once to haul a horse out of swampy ground near Gladstone after it became stuck, pulling it to safety using ropes.

The squad helped out in many other successful searches over the years, including one in November 1986 in the Central Highlands when five Hobart schoolgirls were lost for several days. But others had very different results. In 1977, they scoured the bush as part of an SES search for any trace of the remains of a Tiger Moth plane that had disappeared five years before. The plane had gone missing with Brenda Hean, an environmentalist campaigning against the flooding of Lake Pedder, and her pilot aboard. Nothing was found and the mystery remained. They were called out on occasions to what turned out to be drownings, riding the shorelines of lakes after reports of capsized boats, looking for lifejackets, fishing rods or bodies. On one occasion while on a training exercise in the Lakes country they had to get a canoe to the lake as quickly as possible from 500 metres away. They hoisted the canoe up and onto the rump of a big quarter horse cross, resting it there. The upturned canoe extended several feet either end of the horse, covering it so that the horse couldn't raise its head or see out of it. That horse didn't worry a bit, says Terry. The squad was asked on several occasions to join searches for people who had gone into the bush and committed suicide. Sometimes people who'd been deemed lost walked out of the bush onto a road or to a farmhouse by themselves. One 'missing' man was discovered at home watching television. On many more occasions, the team would be put on standby and not go anywhere at all.

When Bill Bessell, a leader and mate to the men, died in a car accident in 1983, the team decided to push on in his memory and redoubled their efforts, upgrading their skills in map reading, radio communications and rescue procedures. They changed their name to the Northern Tasmanian Mounted Search and Rescue Squad and introduced a uniform—a red jacket with a horse head logo, a shirt, checked pants and an Akubra. They set themselves a target of two hours to get ready after a call-out. That meant getting together a team of six to eight men with their horses, each with provisions for 24 hours, spare clothes, waterproof gear and first-aid equipment in a pack saddle. The members would have a leather bag always packed and ready with essential items including map-reading gear, radio, and waterproof paper and pencil for noting grid references. You left in 'top gear', adrenalin rushing, says Terry.

Always a handy man, Terry designed and built a 'rescue litter'—a canvas stretcher with traces that attached to either side of the horse's saddle at one end and which had a wheel and handles at the other end. One man led the horse while two others held the handles of the litter. This reduced the number needed to carry an injured person to safety, from the eight people carrying a stretcher by hand, in two shifts, to three. The horses were trained in procedures such as pulling the litter but were all sensible types anyhow, good bush horses who did what they were told. There was a 'bit of everything' among them, Terry says, including several Arab crosses, quarter horse crosses and Appaloosas, used at other times mostly for trail riding but also competing in rodeos or at gymkhanas. Terry mostly took out Toby, a paint Arab cross.

In 1986 the squad bought a custom-built 10-metre trailer that could carry eight horses with money raised at rallies and from donations. The trailer had a mobile kitchen in the gooseneck at the front. It was intended to help reduce the costs of transporting the horses—the members had previously funded all the costs involved with the upkeep of the squad themselves, including the maintenance of their floats, horses, feed, care and gear. But the truck proved impractical given that the men in the squad increasingly lived in different areas.

Terry resigned from the group after being president for 25 years in 1998. The squad's now called Tasmanian Mounted Search and Rescue. He was asked recently if he was interested in being presented with a medal for community service in the mounted rescue squad, but declined. 'I was just happy to do it,' he says. That's Terry all over, says his wife Kay. Always helping out, never thinking anything of it, never taking credit and never looking back, unless prompted of course!

Terry, a fit 68-year-old, still rides. He's out in the bush at every opportunity, whether it's taking his children and grandchildren on trails or hunting deer with mates. And, you imagine, that's what he'll be doing for some years to come.

22

A DIFFERENT APPROACH

'That horse is crazy,' said the man delivering the wide-eyed, chestnut colt to Dianne and Mark McIntosh's training centre, shaking his head. 'Good luck.'

Wilbur, a seven-month-old quarter-horse foal, had been booked in at the McIntosh's Horse Workshop in South Gippsland for some early handling and to teach him to lead and tie up. He'd had virtually no contact with humans and had been separated from his mother only that day, run into a cattle yard then loaded onto a stock crate to go to the McIntoshs'.

Mark and Dianne put him in their round yard, where Wilbur paced, sizing up the yard, looking for a way out of it. After a few hours Mark tried to approach the little horse but that was too much pressure for him; Wilbur exploded, springing up vertically over the yard wall. One big leap and he was over the 2-metre high wooden fence and away. Mark and Dianne looked in disbelief and asked themselves what they'd got themselves in for. They'd handled plenty of horses in that yard before and had never seen anything like it.

'In that first week we wondered what on earth we were going to do with him,' says Dianne. 'We didn't want to rush him, he was only a baby and he was petrified of us.'

At the time, a teenager who lived close by had taken an interest in the McIntoshs and their horses. David O'Meara would drop by on his way home from work at the dairy up the road. His sister Daphne had lessons at the centre but David had never been interested in riding. But he was interested in Wilbur. He watched from the viewing platform above the round yard as Mark tried to get the nervous colt to respond to him. Wilbur had escaped from the round yard a second time. 'He still didn't want to know us,' says Mark. 'He wouldn't connect with us.'

Mark had noticed David's curiosity in the colt. David has Asperger's syndrome, which makes him hard to read at times, but his concern for the frightened animal was obvious. 'Would you mind sitting in here with him?' Mark asked him. 'Just sit in the middle of the yard and read a book.'

David nodded without saying anything, looking at the ground, as he usually did. He sat on a chair in the sandy round yard, head down in a book. Wilbur looked at the slight youth sitting so quietly by himself, interested. The foal cautiously approached.

'Why don't you lead him round?' Mark suggested.

Soon, David was leading Wilbur, youth and colt calmly walking side-by-side, in sync. Wilbur's ears flicked as David spoke softly to him. It was a beautiful sight.

David always had an affinity and empathy with animals, his mother Heather says. 'He seems to be able to communicate with them and they just know that he's not going to hurt them. When he saw that foal, he saw that he was scared, didn't have any friends and that no one seemed to understand him.' Animals take things on face value and so does David. If he sees a traffic sign that reads 'Police enforcing speed', he'll interpret it literally. Dianne once told Mark off for asking David to 'hold the horse' for him while he was doing something; David would think that the horse was way too heavy to hold, she pointed out.

At the time that Wilbur was in their care, the McIntoshs had just completed a trainer's course in EFL (Equine Facilitated Learning), a technique that uses groundwork to help improve self-esteem, communication and confidence in the people doing it. As David helped with Wilbur he inadvertently became their first student.

EFL picks up the principles of natural horsemanship, using leadership to guide a horse rather than force. Horses are herd animals that look for a leader to follow who makes them feel safe, taking cues from that leader about how to behave, Mark explains. In EFL they're worked at liberty so you have to earn their respect to get them to follow commands, otherwise they ignore you.

After learning some points about safety, participants are taught how to tell a horse in a yard to walk, trot and canter, turn around, back up, stop and come in to them. Because they're instructing the horse with voice commands and sounds, they learn to speak up and communicate clearly. They learn that horses, being sensitive prey animals, pick up on intent and emotion, and respond instantly to anger, fear, aggression and anxiety, becoming a mirror to the person handling them. A horse can see a bad mood approaching from the other side of the paddock, Mark says, and they won't have a bar of you. They react to body language, gestures and even breathing.

The young people in the course are taught all of this, but they find out themselves what it means to earn a horse's trust and respect—and that can have a profound effect on them.

David changed.

'It's helped him talk more,' says Dianne. 'When he first came he wouldn't say a word to us, he'd just look at the ground, but if you walked away you could hear him talking to the horse.'

So the McIntoshs would sit on the viewing platform, encouraging David to give vocal commands to Dianne's quarter horse, Shimmer.

Dianne recalls: 'We'd say, "You've got to tell the horse to trot on, let him know what's happening." But David was only talking quietly so we'd coax him, "You've got to talk louder because the horse can't hear you." Then after a while he'd talk to us as well. It's helped him read other people a bit better, too.' David also started to talk more when he was out with Dianne and Mark, away from the horses. 'Now you can't shut him up!' she jokes.

EFL is being enlisted for people with autism, bipolar disorder, those who have anger management issues, children with attention deficit disorder, bullies, and victims of bullying, trauma and abuse. It has been used in America and in Europe for many years and is catching on in horse-riding centres throughout Australia.

Mark first found out about it talking to someone at the branch of the Riding for the Disabled Association (RDA), where he and Dianne volunteer. He then heard a radio interview about the 'Horse Boy', the true story of an American boy with autism whose family took him to Mongolia to ride to help his condition. Mark had witnessed the therapeutic benefits of horses at RDA, the immediate changes in the people who rode there, and was interested in the benefits of groundwork. When the couple found out that the American 'horse whisperer' Franklin Levinson, who instructs on EFL, was coming to Australia to conduct clinics, they signed up.

The four-day course was challenging.

'We'd both been around horses a whole lot and this was a bit airy-fairy at first,' says Mark. Mark had grown up with horses all his life and was a no-nonsense paramedic, a 'hard bastard' by his own reckoning. Dianne had ridden since she was four, competed in showjumping and dressage, picked up ribbons at the Royal Melbourne Show twice and had broken in horses, among other experiences. She works as an integration aid in a primary school, and saw in Franklin a real empathy for children with special needs, a knack of linking them and horses together.

Mark was impressed by Franklin's knowledge of horses and how they think. 'He suggested to me when I had a horse in a round yard and was making it go harder, "Why not go softer, relax?" I did and it worked—that's the thing I took away from it.'

The couple went home enthused.

That was in 2010 and the McIntoshs have been teaching classes in EFL ever since. They instruct children, teenagers and young adults supported by Yooralla, the disability services group, and take private students, some of whom have been recommended by a local psychologist.

Sometimes progress is slight. One young man in the Yooralla group was reluctant to even approach a horse at first, and the most Mark could do was persuade him to touch his little finger on the horse's nose. He progressed to leading a pony in some sessions; on others he'd stand stroking a strand of mane Mark cut for him.

Very occasionally they'll get someone who's not suited to EFL at all but Mark and Dianne have otherwise been heartened by what they see happening in the round yard.

'The horses are asked to do simple tasks and they just do it,' Mark says. 'There's something about the way they read these people that's quite magical. And once the participants understand how to ask for and receive cooperation from a 500-kilogram horse, their confidence and self-esteem goes sky-high. As a paramedic you had to be hard to survive—but watching this just blows me away.'

A five-year-old autistic boy with a short attention span was a handful at first, wandering off during his first lesson. The next week Dianne encouraged him to lead a pony over a pole on the ground. 'He yelled out "Jump" when it got near, and then did it again. It mightn't seem much but for that boy it was a big step forward—he concentrated, followed a task and vocalised a command.'

The Yooralla carers say they have noticed changes in the behaviour of the people they take by minibus for their weekly EFL class. One young woman, who arrived timid and softly spoken, gained confidence as she noticed the horse's ears pricking up when she gave it a command. She started speaking up, projecting her voice clearly and, according to the carer who came with her, her confidence continued outside the lessons.

David O'Meara was well suited to doing EFL and working with horses. He doesn't look at you when he's talking, unless he knows you well, and horses instinctively respond to that—as prey animals they get nervous when they're being watched. Perhaps his biggest triumph during his informal EFL training came the day David persuaded Shimmer to lie down, commanding the big chestnut to drop to his knees then over onto his side—quite an achievement as this puts the horse at its most vulnerable. These aren't trick ponies, says Mark; they have to trust you to do that. David just shrugged his shoulders afterwards and said, 'Oh, yeah.' His mother Heather, watching outside the yard through a chink in the gate, was in tears. She told the McIntoshs later that her son talked about it for weeks. 'There's something about seeing your child succeed at something like that. I could see in David's face that he understood the horse trusted him.'

When Wilbur went home after the McIntoshs had taught him some basic handling, David, who rarely shows emotion, said, 'I'm going to get my own horse'. Two years later he did, and although he'd never ridden before he broke her in.

Holly was a two-year-old chestnut quarter horse who'd had a bit of halter training at most. Daphne bought an unbroken, purebred quarter horse, Ted, from the same owner, and both horses were delivered to the McIntoshs to be started. (Like others who practise various forms of natural horsemanship, the McIntoshs prefer to say they 'start' rather than 'break in' a horse.) The first thing the two horses did when they arrived was to run off to the other end of the paddock and refuse to be caught. 'She did a runner,' says David. 'I could've sold her the first day I got her.'

David had learned how to safely handle horses through the EFL instruction and went on from there. Holly proved to be strong-willed and liked to do things her way. David understood that instantly, knowing when to push her and when to back off. 'She's stubborn,' he says, 'but she's a bluffer.'

After some groundwork in the round yard, and practice in desensitising the horses, the McIntoshs introduced a bit and bridle, without the reins, and a saddle pad. After a few days they tightened the bit. Everything was done slowly, gradually. David learned how to drive Holly in long reins. They put an old saddle on her with a sandbag, getting her used to the idea of weight on her back, then shifted it sideways so that if David ever started to come off she wouldn't startle. They slapped the saddle so she'd get used to flapping knees. Then one day Mark jokingly told David to get on her, and he did. David didn't understand that Mark was joking and didn't think to question him anyhow—he knew Holly was ready. She didn't flinch.

Not one to show excitement, he casually mentioned it to his mother that night. 'Yeah, she let me hop on her today then I hopped off.'

David developed his own idiosyncratic ways of handling his horse once he got Holly home. Whereas Daphne, who wants to compete in reining and dressage, is more demanding on Ted, knowing what she wants to achieve with him, David says Holly's already 'good' and likes her as she is.

'They've got a bond, an understanding,' says Heather. 'David looks at Holly when she's doing something "wrong" and says, "She's just being a horse, that's what they do". He's happy for her to just be a horse. He's not a forceful leader but he's her leader.' Holly sticks to him like glue. She tends to kick out at other people but not at David.

She can, however, be a 'doofus' sometimes, he says, and they still have 'arguments' when she baulks at doing something and stamps her foot. David once smacked her over something she'd done; Holly bit him back hard, so David gave her another smack then ran away before she could bite him again. 'She was fine after that, I made the point.'

He looks out for her, too, and became very worried when Holly appeared to be choking on her dry feed on a couple of occasions. Dianne suggested wetting her feed with water. David took the advice literally, adding so much water that it was like soup. 'She likes it like that!' he says.

David does whatever works for him and Holly. He decided early on that he didn't like trotting because he's thin and found it uncomfortable on his bottom, and resisted being taught to rise to the trot. He's happy to walk Holly. If he's trail riding with the others and they speed up, Holly will ignore the other horses and stay at a walk, sensing that's what David wants. They're doing it their way.

'We're a team,' he says, looking up.

PART VII

Tales of the unexpected

23

THE HORSE ON THE ROAD

The weekend started normally enough. It was the Saturday of the Australia Day break in January 2001. Rod, a journalist working in Melbourne, was driving north along the Hume Highway headed for the Kiewa Valley, where he would stay with a friend on her sister's farm. He was looking forward to it. He hadn't seen his friend Nell for a while and it was good to be out driving in the country.

As he drove he thought contentedly about the coming visit, not knowing that an encounter ahead would make that weekend stand out in his mind forever.

He passed the turn-off to Avenel, where he'd spent his early boyhood. The family had lived on a sheep farm several kilometres outside the town and he had gone to the tiny three-classroom school there. Avenel is a pretty place, a settlement overhung by shady trees by the Hughes Creek. It has the distinction of having been home to Ned Kelly when he was a boy. Rod grew up hearing stories about the Kellys, such as the story about the time Ned, who had gone to that same school, saved a young boy, Richard Shelton, from drowning in Hughes Creek. Richard became the grandfather of former Essendon footballer Ian 'Bluey' Shelton, one of several footballers in the clan. Richard's family gave the eleven-year-old Ned a green silk sash as a reward, which Ned wore as a cummerbund at the Siege of Glenrowan where his outlaw gang ended its days. Rod remembers a practical joke in which his sister, in a grade above him at the composite school, said she had 'found' an old pencil on the classroom floor with 'N. Kelly' written on it.

The area Rod was driving through was surrounded by Kelly country. Rod had learned to ride on the granite slopes so characteristic of the area but the family sold the farm in 1962, when he was seven, and moved to Melbourne, taking their horses with them. He'd kept riding and in his early twenties worked as a jackaroo on stations in the Northern Territory. There had been many horses in his life but never one the likes of which he would soon see.

Rod turned off the Hume near the town of Glenrowan and headed in the direction of Myrtleford, passing through the small town of Oxley. He stopped at Milawa to visit Brown Brothers' winery and buy a bottle or two for his hosts, and then backtracked to Oxley, to investigate a 'foodie' store he'd noticed on the way through.

It was about one o'clock when he left Oxley, driving on the Glenrowan–Myrtleford Road towards Milawa again. He passed the last of the town's houses on a long curve of road that veered to the left. The land on either side of the road opened up into flat paddocks, yellowed by summer and with wide-girthed red gums spread sparsely throughout them.

Something to the right caught his eye. A chestnut horse with a white blaze was galloping across the paddock. Fast. On its own, it was moving at break-neck speed towards the road. It was unusual to see a horse move like that of its own volition, Rod thought. There was nothing chasing it and no other horses in the paddock.

The horse was perhaps 200 metres ahead of him, approaching the fence and the road. Still driving slowly after leaving the town, Rod expected to see it stop suddenly at the fence. But it didn't. The horse kept galloping straight across the bitumen. He looked at where it had gone, expecting to see a lane or driveway on the other side of the road, somewhere it had bolted down—because it hadn't stopped at the fence on the verge on the other side or jumped it. He drove to where he'd seen the horse cross the road and pulled over. There was no lane or driveway, just fence. He looked at the wide-open paddock beyond it. It was empty. The horse had vanished.

Rod paused to think about what he'd just seen. Checked the paddocks on either side of the road again. Still nothing. Eventually he drove on, replaying in his mind what he had witnessed. It occurred to him that he hadn't seen the horse jump either fence. It had passed through them. There'd been cars coming in the other direction and Rod wondered whether the drivers in those cars had seen anything. It was a fine day with clear visibility. What he'd seen didn't make sense, but he was fascinated rather than frightened by it. 'I didn't resist the idea,' he says. 'I just thought I'd seen something that wasn't there. It was a paranormal experience.'

Rod told Nell about the sighting of the horse when he arrived at the farm, laughing about it with her. She had little to say beyond that it was 'amazing', nothing that she could add that would make sense of the strange occurrence.

When he returned home from the weekend he recorded the event in a diary. It reads:

> On a sweeping bend leaving the town I was surprised to see a chestnut horse with a white blaze gallop out of a paddock and across the road, mane and tail streaming. I thought it had got free and was galloping down a lane as it did not pause at the paddock fence as I expected. I was amazed and thought it was lucky not to have struck a car. When I got to where the horse had been, I saw there was no lane, just barbed wire fences on either side of the road. There was no horse to be seen anywhere.

Rod thought little about the experience afterwards until a couple of years later when he was reading some material about the Kelly gang. He'd had a passing interest in Ned Kelly since childhood; the story of the outlaws and the siege at Glenrowan had, after all, permeated the psyche of the small town of Avenel, like others in the area. Ned Kelly's father and brother are buried in the cemetery there.

Five decades later he felt a shiver run through him as he read a passage about the gang member Joe Byrne and his horse. 'The reference was to Joe Byrne and some death-defying ride he'd done to bring people a message or get away from the police or similar,' Rod recalls. 'Then I read that people had reported seeing his horse in that Oxley area without a rider. It was a phenomenon that people reported seeing over many years. I thought, "Oh shit, I've seen Joe Byrne's horse's ghost."'

Joe Byrne is noted in history as a skilled horseman, a daring rider who would ride his horse down steep gullies flat-out just for the thrill of it. He, like others in the gang, bought, borrowed, stole and sold horses all around the district. Little is noted about his horses other than that he often shared Ned Kelly's horse, Music, a grey mare who positioned herself between the outlaw leader and the police as fighting raged during the Siege of Glenrowan. Joe Byrne was fatally wounded on that day, 28 June 1880. The horse he'd brought with him was later discovered tied up in a stable behind Macdonnell's Railway Tavern nearby. It had been one of two horses that had been 'borrowed' a few days before from their owners in Dookie and was photographed being held by police after the siege. The photo is of course in black and white—but the horse in it has a broad blaze, and it was reported that its colour was chestnut.

24

SURPRISE MOMENTS FOR AN EQUINE VET

Julian Willmore graduated from the University of Queensland in 1975, began work as a vet in a mixed rural practice then moved into equine work. He now often treats horses at the elite level of horse sport. He has vetted with the FEI (Fédération Equestre Internationale), worked at various World Equestrian Games, and the 2000 and 2008 Olympics. He recalls some of the more unusual cases in his busy equine-only practice in southeast Queensland.

I don't know why horses get into empty concrete swimming pools, but they do. Maybe they're grazing around the house at night and escape their yard, push through the fence to go after the succulent lawn around the pool, alert the family dog, scramble across the concrete pool edge and fall down inside.

One incident involved a pony that got stuck in a pool which had been emptied of water. The pony was standing there gazing up at the steep sides when I arrived. The owner was a civil engineer and decided he would build a timber ramp for the pony to walk up. My role was just to check the bloody scrapes on the pony's legs—he was doing the rest, he said. He wanted to know how heavy the animal was to be sure that his ramp would hold up, then said my help wasn't needed anymore.

As with most horse-driven families, the wife's demands were met and I was asked to return. The pony was still in the pool the next day with hay and manure strewn around the pool floor. The ramp was nowhere near suitable, so I gave the pony a quick anaesthetic and fixed some shackles to its lower limbs, and it was lifted out sound asleep using a neighbour's backhoe. It recovered uneventfully on the grass poolside.

Lifting horses by their lower limbs while under anaesthesia is an acceptable method of moving the animal in most surgeries to and from anaesthetic induction stalls to equine operating tables. Another horse under my care was lifted out of a pool in the same way.

I've often had Shetlands get into feed sheds. The ponies are so hungry by nature that they find their way into chicken coops too, eat all the chicken pellets, push the door shut behind them, can't get out and suffer from a stomach grain overload. They push the chicken wire out of shape, knock over all the water containers and feed trays, then stand on most of them, bending them or destroying them as they're waiting to be rescued. They usually need the vet to administer colic treatment after the owner has tried beer and other bush remedies such as Worcestershire sauce poured straight out of the bottle into the horse's mouth, probably resulting in more sauce ending up on the ground and over the handler than in the horse.

I was once contacted by the owner of a horse who was distressed that their horse's manure was red, assuming it was blood. I examined the horse and the manure and reassured them that the horse wasn't passing blood. Then we had to work out what the horse was eating that would turn its manure red.

As it turned out, the horse lived near a food-canning factory. The waste beetroots from this factory were used by a farmer to feed his cattle, as many farmers source waste food. The horse did not normally have access to the cattle feed but apparently had made its way into the same paddock to feed on the big pile of dumped beets that were unloaded in bulk into the field for the cattle. The horse ate the beetroot and the red vegetable came through the horse as red-coloured manure. Mystery solved.

While I was official vet at the Brisbane Royal Show—for sixteen consecutive years—my role was really dealing with equines. There was a cattle vet, other vets for the dog show and so on, at least so I thought until a government vet approached me about checking the health of an elephant. The elephant was at the attached zoo, which was part of the Brisbane Show attraction. The government vet said it was not a government job but a private practitioner's job. I was young and inexperienced; he was older and wiser and knew how to avoid any scary procedures but apparently it was my job. The elephant was of concern because of a suspected tuberculosis lump between its front legs on its underbelly so it was my task to collect a sample from the elephant. The handler was of some help. He had the elephant stand on a small drum with all four legs positioned on the drum then one leg uplifted to keep the animal in place. But this didn't stop the elephant's trunk from being airborne and the elephant hitting me a few times with it in between trumpeting out loudly. The noise stirred up the circus dogs that were chained nearby, causing them to become so excited that they ran to the length of their chains, barking at the monkeys chained next to them. The monkeys were just within reach of one of the dogs, which bit one of the monkey's fingers, taking off the fingertip. The handler, having finished with the elephant, told me the monkey deserved the bite because the monkeys aggravate the dogs by constantly throwing elephant faeces at them.

Now I had another headache, more unfamiliar territory—I had to treat an injured monkey's finger! The handler held the monkey for an examination but it behaved like a squirming child, was aggressive, glared at me and showed me its teeth as a warning to keep back. I was then wishing I'd never been involved with this entire fiasco. The handler told me to befriend the monkey and feed him some shredded salad mix but as soon as the monkey took the food bribe he threw it back at me. I decided his finger wasn't too bad and didn't warrant me having mine bitten off in an exchange so I abandoned the job, returned to the horse stables, pleased to say that zoo practice wasn't for me. I've stuck to horses ever since!

25

BEWARE THE FREE HORSE

It seemed like a good idea at the time. Andrew of Oodnadatta was picking up a truck and a few horses in Queensland to transport back to South Australia and the owner turned up with a couple of extras he could have for nothing. Why not? he thought.

He'd planned the trip well. Andrew was busy running two farms with his wife and couldn't afford to stay away for too long so had the whole journey organised. He was going to a property near Dalby on the Darling Downs to buy a gooseneck trailer, which he'd connect to his prime mover and bring home. Before he left he cast around for anyone wanting horses transported from Queensland and ended up with an order for six, which he arranged to be delivered to the property of the people selling him the trailer.

He would drive 2500 kilometres from Oodnadatta to Queensland over two days, pick up the trailer, load the horses early the next day and drive back, breaking up the journey with an overnight stay in Dubbo, New South Wales.

The first half of the trip went according to schedule. The two-day journey south towards Port Augusta, east towards Dubbo through New South Wales, then north, was pretty rugged, the roughest couple of thousand kilometres Andrew had ever driven, but he got there on time and without event.

Andrew collected the gooseneck trailer that night, and on cue the horses to be transported arrived at where he was staying. He was slightly surprised, though, when the woman who delivered them mentioned there were two extra horses he could have, but they looked fine so he agreed to take them. The horses all loaded well early the next morning and Andrew headed down the Moonie Highway towards Dubbo, 700 kilometres or so away, happy that he'd made a good start. He made it to Dubbo in reasonable time, let the horses out at the local pony club grounds, fed and watered them, and went to sleep contented. The trailer with its kitchen and sleeping area seemed like a dream come true.

Andrew awoke before five o'clock the next morning and fed and watered the horses. One stop at a fuel station he remembered as being just out of town and they'd be on their way. His destination was Hay, New South Wales, for a rest, then on to Oodnadatta.

He began loading the horses at six o'clock, starting with one of the free horses, a young bay stock horse. But the youngster didn't want to get back on the truck. He'd stood in the trailer for seven hours the day before and didn't like the idea of doing it again. Andrew took him aside and loaded the rest of the horses. All popped on beautifully. Then he went back to the little horse, led him round a bit, played with him, no pressure, then tried him again. The horse wasn't having a bar of the trailer. Andrew became firmer, the horse pulled back on his halter and stood his ground, looking at him. An hour later the other horses were still standing quietly in the trailer and the young horse was still mucking around. He wasn't scared; he just knew he didn't want to get on again.

Andrew hadn't brought much gear or tack with him and had no one to help him push the horse up. He decided to try another method, leading the horse to the ramp then lunging him for a while every time he refused to walk up it. The horse kept refusing and Andrew kept lunging, hoping that if he gave him a bit of grief he'd tire and give in. He didn't.

Andrew took the lead rope off another horse in the trailer and tied it round one of the bay horse's front legs, pulling him by the halter and the leg at the same time. That got him halfway up the ramp. He took another rope off one of the other horses, tied it round the other front leg and pulled both legs. The horse strained back on the ropes as far as he could, sprawling out along the ramp, belly nearly touching it. Andrew got two more lead ropes and tied them round the horse's back legs, leaning back at the top of the ramp, straining to budge him. Andrew's a 'fairly big fella' by his own reckoning, 6 feet 2 inches tall, but the little horse was stronger. Andrew wedged his heels into the cleats on the ramp, put the lead rope in his mouth and pulled on the four other ropes, and managed to get the horse to shift his feet one at a time, by now feeling like he was in some sort of *Footrot Flats* cartoon. He was too wrapped up in what he was doing to notice if anyone was watching the spectacle.

As the horse inched closer to him up the ramp Andrew suddenly wondered how he was going to get out of his way once he got to the top but he couldn't stop now—it had taken most of the morning to get the horse this far. As he wondered, the horse reached the top and jumped straight over him and into the trailer.

Gotcha! Andrew tied him up and latched up the ramp, cursing the horse he realised he knew nothing about. It was eleven o'clock—five hours after he'd started loading.

With a twelve-hour drive ahead of them, Andrew decided to go straight home. It was a solid kind of drive for the horses but he couldn't risk letting them out for a break at Hay and having to get the free horse back onboard.

Next stop was the fuel station on the outskirts of Dubbo. By now flustered and hot, Andrew noticed that the first fuel station was full of trucks. Keen to keep moving he decided not to wait there and drove past it. The second fuel station didn't look like it would fit the prime mover and trailer so he bypassed that one, too. But as he reached the end of town he realised there were no more fuel stops. He put the indicator on, turned into a street to backtrack to the first stop—and ran out of fuel. It was all the free horse's fault! He swore as he parked, cursing it, anyone and everyone and especially the woman who gave him the horse.

He got out and found a go-kart shop, bought a fuel container for petrol and walked back to the second fuel station, paying a woman there to drive him back to the truck. By the time he'd gone back and filled up it was 12.15, and he was running more than six hours late.

Andrew drove through the afternoon and night, checking the horses regularly, cursing the free horse every time he looked at him. It was two o'clock in the morning when he got home. He unloaded the horses and was pleased to see that they all walked off the trailer as 'fresh as daisies'.

Andrew came to terms later with the free horse, an unbroken stock horse. He says he was a polite little horse that didn't kick or bite. A horse-breaker that Andrew had transported horses for on that run agreed to break in the horse in exchange. Then there was the matter of what to do with him.

The woman who'd given Andrew the horse had suggested at the time that he might make a good kid's horse, but Andrew and his wife breed competition horses and their son had the pick of the crop, so they decided to move him on. Andrew rode the free horse after they made the decision and almost wavered, but he had a client who wanted a horse and the freebie, who'd grown into a good-looking animal, fitted the bill. They sold him to her for a reasonable price. She sold him on and Andrew doesn't like to think where he is now. He thinks it's a bit sad that there are so many others like the free horse around.

They've still the got the other free horse from that trip, though. He's called BRB, short for Big Red Bastard. He bucks a lot.

PART VIII

Hanging on to our heritage

26

THE *REAL* GEEBUNG POLO CLUB

It started in fine tradition: a few blokes were talking over a beer or two at the pub when the subject of bush poetry came up, and a favourite poem, Banjo Paterson's *Geebung Polo Club.* Mountain cattleman Ken Connley and Phil Macguire and their mates were remembering aloud the lines of the poem about the grudge game of polo between the bush riders on the Geebung team and the city toffs on the other when someone suggested forming a team in honour of the famous poem.

That was in 1989 in the Dinner Plain hotel and the annual Geebung Polo Club game has been going strong ever since. The first game was a bit of fun, says Craig Orchard, who's played in the bush team from the start. A 'scratch match' was held with local riders in the hotel's carpark, in the snow, using trees as goal posts. Wasn't easy playing in the snow, says Craig. About 30 people stood watching.

The following year the match was held at Horsehair Plain, where the Hotham airport is now. A 'Cuff and Collar' city team came up from Melbourne. The bush versus city rivalry began in earnest.

The city team's numbers include some of Australia's best, names like Robbie Ballard and Jeremy Bayard, who play on the polo circuit here and overseas. They occasionally turn up with an international player, too, perhaps someone from England or South Africa. A few years ago they brought along an Englishman, Lord Tim, fifth in line to the throne (or so it was rumoured) and a chap who played polo for a living. Lord Tim could belt the ball from one end of the field to the other. Had three shots at goal and got them all.

The Geebung side started out with ten or so players who would swap around between chukkas (like quarters). But Ken Connley, 'Man From Snowy River' stuntman and the captain, got rid of a few of the 'useless buggers' and pared the team down to include himself, his late brother Rusty, Rusty's son Joe, who's had offers to play competitively, Craig Orchard, and David Olsson, who's represented Victoria at national level playing polocrosse.

Joe Connley is the best hitter, Craig backs him a lot, Ken always rides on the other team's best players and, in his words again, 'roughs them up a bit', and David plays all over the ground.

The Geebungs get together to practise only once or twice a year, though they meet at Lake Omeo to discuss strategies over a few beers, too. 'Captain Ken doesn't want us to peak too early,' jokes David. 'We're getting better every year, though, more skilled,' he adds.

'We just have to play a bit rougher or dirtier than them,' says Craig.

The annual Geebung match is played under Banjo Paterson's rules—that there are no rules.

'If they were beating us playing properly, Captain Ken would rope a rider from the other team or crack a stockwhip,' says Craig.

Ken admits to letting the other side's horses loose during a rest between chukkas and is quite happy for the Geebung supporters to widen the goal posts at their end for them. One of the more serious players on the city side, the late Bryce Dicks, sometimes got upset at the lack of decorum. They've brought in a rule or two now that they're older and wiser, says Craig. Like not galloping into a player who's about to strike the ball with the mallet. Otherwise they can clock you and it hurts. There's the occasional fall, blood and bruises, and the players sometimes nearly come to blows, but that's as close as they get to the biffo and bloodshed of the bard's poem. 'Nothing that a night at the Dinner Plain hotel can't fix or a few days limping afterwards,' says Craig.

Neither team has dominated over the years.

'The big polo players think they should beat us every time, but we can outride them with our horse skills and bush skills,' he says. The Geebungs gave the city team a flogging when they met in Easter 2012, but lost by a goal in 2013.

The game traditionally starts with a recital of the Banjo Paterson poem before the ball is tossed in from the sidelines by the umpire and the mallets start whirling. There are four players a side, who can swap horses or riders between chukkas. The flat-out galloping is hard on the horses and riders, says Craig, runs the horses into the ground. They have a few beers between chukkas and something to eat in middle.

The local riders bring their own horses, which they use on the farm, at rodeos or for brumby running. Ken Connley supplies the city team with horses, admittedly less polished than the 'natty little ponies that were nice, and smooth, and sleek' that the toffs rode in the poem. The Melbourne fellows complained once that the horses they were given to ride weren't up to scratch and brought their own horses the following year. 'We *really* beat them then!' says Craig.

The teams are easy to identify. The Cuff and Collar team are the ones in the smart white polo shirts with numbers and belted pants, big leather shin guards and boots and helmets; the Geebungs wear jeans, shirts or T-shirts, bush hats or even a baseball cap and hide their motocross knee pads under their jeans.

A position on the Geebung Polo Club is coveted. Every now and then a young bloke wants to join but Captain Ken says someone has to die before they'll let in a new member.

For many years the match was held on Horsehair Plain but it moved more recently to Cobungra Station. 'The field there's real good, a bit similar to a proper polo field—nice and green, ploughed up and mown down,' says Craig. 'The one at Horsehair Plain had snow grass which slows you down, tussocks and rocks, though the rocks were a big incentive to stay on your horse!'

The match is held at Easter and has grown into a big tourist attraction attended by 2500 to 4000 people. It's an all-day event with games for the spectators, lolly drops by helicopters for the kids, chicken-throwing competitions (using frozen chickens), marquees, bands, and wining and dining. The toffs from the city sit in their convertibles sipping champagne or take a break from the chukkas for a bite in the big marquees; the locals perch on hay bales or sit on the tailgates of trucks or floats, drinking beer. Or something like that.

At the end of the game the players and their partners get together for a drink at the pub and a presentation ceremony. The big Geebung Cup is presented to the captain of the winning team, the Rusty Conley Memorial trophy to the best player on the ground, and the Bryce Dicks Memorial Trophy to the most determined. Then the city folk retire to their swish chalets and the country folk bed down for the night in swags outside the pub.

The Geebung team has been invited to Melbourne to play but 'you know, we don't want to get too good!' says Craig.

And, 'We love the mountains, mate, we've got farms to run,' says Ken. It just wouldn't be the same.

The Geebung Polo Club by A. B. 'Banjo' Paterson

It was somewhere up the country in a land of rock and scrub,
That they formed an institution called the Geebung Polo Club.
They were long and wiry natives of the rugged mountainside,

And the horse was never saddled that the Geebungs couldn't ride;
But their style of playing polo was irregular and rash—
They had mighty little science, but a mighty lot of dash:
And they played on mountain ponies that were muscular and strong,
Though their coats were quite unpolished, and their manes and tails were long.
And they used to train those ponies wheeling cattle in the scrub:
They were demons, were the members of the Geebung Polo Club.

It was somewhere down the country, in a city's smoke and steam,
That a polo club existed, called the Cuff and Collar Team.
As a social institution 'twas a marvellous success,
For the members were distinguished by exclusiveness and dress.
They had natty little ponies that were nice, and smooth, and sleek,
For their cultivated owners only rode 'em once a week.
So they started up the country in pursuit of sport and fame,
For they meant to show the Geebungs how they ought to play the game;
And they took their valets with them—just to give their boots a rub
Ere they started operations on the Geebung Polo Club.

Now my readers can imagine how the contest ebbed and flowed,
When the Geebung boys got going it was time to clear the road;
And the game was so terrific that ere half the time was gone
A spectator's leg was broken—just from merely looking on.
For they waddied one another till the plain was strewn with dead,
While the score was kept so even that they neither got ahead.
And the Cuff and Collar captain, when he tumbled off to die,
Was the last surviving player—so the game was called a tie.
Then the captain of the Geebungs raised him slowly from the ground,

Though his wounds were mostly mortal, yet he fiercely gazed around;
There was no one to oppose him—all the rest were in a trance,
So he scrambled on his pony for his last expiring chance,
For he meant to make an effort to get victory to his side;
So he struck at goal—and missed it—then he tumbled off and died.

By the old Campaspe River, where the breezes shake the grass,
There's a row of little gravestones that the stockmen never pass,
For they bear a crude inscription saying, 'Stranger, drop a tear,
For the Cuff and Collar players and the Geebung boys lie here.'
And on misty moonlit evenings, while the dingoes howl around,
You can see their shadows flitting down that phantom polo ground;
You can hear the loud collisions as the flying players meet,
And the rattle of the mallets, and the rush of ponies' feet,
Till the terrified spectator rides like blazes to the pub—
He's been haunted by the spectres of the Geebung Polo Club.
The Antipodean, 1893

High Country Polo by David Olsson, Geebung polo player

There is a game of Polo,
Played every Easter Sunday morn,
It was way back in the eighties,
That the idea was first born.
The country versed the city,
Upon the snow grass lawn.
The players displayed their finest skills,
And the ladies hearts were torn.

The Cuff'n'Collar boys,
Were tough and keen and mean,

Cornelia Selover, co-founder of Tuk-Tuk 100% Horse clothing company, found a new lease of life after the Black Saturday bushfires in 2009.

Neville Gilpin (right, pictured with friend Merv Allen) says it's hard to keep Clydesdales in work but he's happy to do it.

Clydesdale fan Merv Allen is keeping alive the traditions of harness by hiring out his wagon and horse, and still harrows soil on his Leongatha farm by horsepower.

The toffs taking on the mountain men in the Geebung polo match at Cobungra Station, Victorian high country.

That's the spirit! Geebung polo matches allow leeway for a certain amount of rough play.

Geebung polo match, players Craig Taylor (L) and Ken Connley (R).

PICTURES: CHARLIE BROWN – MOUNTAINSIDE (WWW.MOUNTAINSIDE.NET.AU).

Clancy, five days after he was rescued in August 2009.

Maddie riding Clancy, left, with her sister Kate and father Tom. Victorian high plains.

Wayne Armstrong with Trooper, the once wild Waler he's training as part of the Mitavite Equitana Waler program, shortly after he arrived. PICTURE: KERRY "PINKY" COLLINS

Helen Packer takes some time out with one of her trail riding horses at her Anglers Rest property.

Leslie Langtip, World War I infantry soldier was in the famous charge of Beersheba in 1917.

Beth Mackay leading Gypsy at the family farm at Woolamai in Gippsland.

Wayne Armstrong competing with his four-in-hand team at the Victorian Combined Driving Championships, March 2013. He won.

PICTURE: TONY JAMES.

Paralympian Anne Skinner puts Cossack through his paces in the dressage arena.

Anne Skinner with Cossack, the horse who, like Anne, is lucky to be alive, December 2007. PICTURE: *HERALD-SUN*, MANUELA CIFRA

Equine vet Julian Willmore treats everything from elite equine athletes to ponies stuck in swimming pools.

The author and Poppy, a small horse with a lot of attitude. PICTURE: ROD MYER

Grae Payne and Samson, the horse that chose him.

Helen Packer with Spook, (both right) and guests at her Victorian high country retreat.

Zelie Bullen with her beloved Bullet.

PICTURE: NAZ MULLA WWW.NAZMULLA.COM

Zelie Bullen performing on Bullet.

There was Jim Castricum, Jeremy Bayard,
And Jim Nolan on the scene,
With Rob and Greg as captain,
The boys played as a team,
Adding Craig, Geoff and Bryce,
It just added to the cream.

The Geebung boys were cunning,
But upon the field did flow,
They matched their rivals with bushmens tricks,
And gave the crowd a show.
With Mad Jack and Hat and Catty,
And Rusty, Scoof and Joe,
Jock, Dean, Craig and Backman,
Even Olsson had a go.

Ken, the Man from Snowy River,
Came down to lend a hand,
He also liked to sing a song,
And never needed a band.
Old Ace was there and Husky,
When the first few games were run,
A lot of people tried to catch him,
But his race was always won.

Rusty and his grey horse,
Were the quickest on the field,
And when the band struck up at night,
Old Rusty wouldn't yield.
He would dance the girls till daylight,
If they could last that long,
He must have had good leather soles,
For he never missed a song.

Matched on the ground as equals,
Rode Pretty Boy Craig and Joe,
And after the game was over,
They continued to put on a show.
In the evening lamp light,
Full of beer and scotch and coke,
They tried to woo the girls over,
Just to see who'd get a stroke.

By now the game will be over,
And we will know who's won and lost,
I would like to toast the players,
On their winning and their loss.
May the game be played for years to come,
Through the good times and the bad,
And we all continue playing,
As though we're just a lad!

2008

Down on the lake on Friday night,
The Geebung boys did meet,
They hit a few balls around,
And thought they were pretty neat.
They had the Bryce Dicks trophy hanging on the wall,
And wondered what he would have been thinking as they hit along the ball.
A beer was had to wind things up,
And a quiet toast to Bryce,
Then all went their separate ways,
To settle for the night.

The Cuff'n'Collar boys would be up at Dinner Plain,
Trying to work out their tactics and trying to hide their pain.

For Bryce, their captain, would not be there,
And never will again,
But his legacy will live on,
With how he played the game.
He would not spare his team all day,
And drive them to defend,
In this he would not give in,
Till the very end.
The horses that were provided sometimes,
Did not fit the task,
But this didn't hinder Bryce,
Who could hit goals out his arse.

Now the game of Polo is over for another year,
Both teams are gathered round the bar,
And the winners let out a cheer.
I do hope it is the Geebung boys,
Who have won it back to back,
But if the Cuff'n'Collar boys have got up,
We must have been too slack.
On finishing up this story,
There is only one thing to say.
We will all be back again next year,
And gather to the fray.

2010

21 years may seem a long time
But it isn't that long at all
When we ran the first polo match
We couldn't even hit the ball.
Once we worked out
The way to hold the stick

To hit the ball on the side
We caught on pretty quick.

Ken is still our captain
For it is he they come to see
It doesn't really matter
If he has a bandage on his knee.
The girls still like his photo
Hanging on their wall
It wouldn't even matter
If he never hit the ball.

Pretty Boy Craig and Joe
Still put it to the test
It is a competition between them
To find out who's the best.
Olsson and Orchard still front up
To play the game each year
The goal posts seem to shift sometimes
When they get into top gear.

We tried to contact the city boys
To see who's playing this year
They didn't even return our calls
Perhaps it was through fear.
They kept their team and tactics
Very close to their chest
They always try and surprise us
And bring along the best.

They two up and coming Geebung players
Ride boundary every year

Hoping they will get a run
Perhaps it will have been this year.
One of us will have to die
For them to get a full run
Frankly boys don't raise your hopes
Cause we're having too much fun.

27

HOLDING ON TO THE HEAVY HORSES

Rob Campbell is yarning, telling us about the time Boxer the draught horse bolted and parted company with the jinker and the family on the way to town one day. Rob and his mother went straight out over the back of the jinker, his siblings somehow stayed aboard. It was 1945 and Rob was only two at the time but he's heard the story so many times it's like he remembers it himself. The family were going shopping in Leongatha, the nearest town to their farm at Berry's Creek, when Boxer shied, broke the traces and bolted on O'Loughlin's Flat.

It must have been a Thursday, says his mate Merv Allen, also sitting at the country kitchen table reminiscing about heavy horses. Because that was shopping day in Leongatha then, the day when the produce was brought to town and the women of the district had the use of the family's cart horse and jinker. The jinkers were open so the women drivers would always carry an umbrella and had a jinker rug for their knees: canvas lined with wool in the inside. The next Thursday the men would go to town to buy cattle. But the women's Thursdays were also the days the pigs were sold so Thursdays were known as 'pigs' and women's day', says Merv. Rob says his father would put the pigs in a spring cart pulled by

Boxer with a net over the top of them; the cattle were driven on horseback. There wasn't a road out of this town that he didn't drove on, he says.

But back to the story . . .

The three children stayed by the side of the road while Rob's mother went for her husband, who came back with some wire, reattached the traces, put Boxer back into them and they continued on as if nothing had happened. That's how it was then, says Rob. Running a horse and carriage was like owning a car, it was all very matter of fact. As was the time that he, his father and sister Dorothy took the sled drawn by a pair of draught horses to pick up the mail and ran over Dorothy's leg with it. The steel runners went over her leg just below the knee. 'But we just got back on it again.'

Rob and Merv were both born in 1943 and just clipped the era when owning a Clydesdale or another heavy horse was as unremarkable as owning a car today. Every family had one. Now, though, they're a little in awe of the huge horses and the parts they played in their lives.

The men grew up on farms outside Leongatha, in the heart of Victoria's dairy belt, southeast of Melbourne. Tractors were just coming in and cars were around, though World War II meant most people delayed buying them. Rob's father had ordered a big American De Soto but wasn't able to get it so the family used the jinker instead. As children, they helped their fathers with the farm work from a young age. Rob recalls how his father would come into his bedroom while he was slumbering early in the morning and bark, 'Get on that horse and take that cow up!' Rob would know he meant a 'chopper' cow, a dairy cow 'that was history', that had to go to the Leongatha stockyards. 'So I'd get on my horse, take the cow up over the railway line, up over the highway to the yards, canter the horse home, get on my bike and get to school. No saddle, no nothing, dirty legs but it didn't matter. No one cared!'

Most families had a couple of acres to grow potatoes and had a horsedrawn potato scuffler that churned through the weeds between the rows of potatoes. A fitting called a 'hiller' would push the soil up against the potato stalks so that the growing potatoes wouldn't be exposed and become ruined. The trick was to walk beside the furrow without letting the horse tread in it and crush any potatoes. Rob recalls having to hold the bridle of a mare who'd never pulled a scuffler before and who kept treading on potatoes being 'hilled' or scuffled. His father would yell out, 'Get her over!' whenever she trod on a spud and give Rob a serve about it. But that mare learnt how to step along the sides of the furrows without going in them. They're smart horses, those Clydesdales, the men agree. Merv operated the potato scuffler, too, from the age of ten, holding on to a handle on the ground while his younger brother sat on the horse steering it at the end of each row.

Most of the farmers trained their horses themselves, teaching an inexperienced horse to lead and tie up, then putting it in with a mature team for a day. Rob's father would 'mouth' the horses, getting them to accept a bit, and teach them to lead in an old dairy yard, in winter up to the horse's knees in mud. After being mouthed, the new horse would be wedged in among others to learn the ropes. They usually understood what was expected of them after a day—but a working day was long then so the new horse had a fair while to get used to it, Rob says. The horses would spend eight hours a day working, five days a week.

They were hitched up in pairs to 'finger mowers' to cut hay, which was left to dry in the paddock then raked with a 'dump rake' that drew it along, lifting it up then dropping it so that it formed a windrow. The men would then use pitchforks to shape the hay into a stack. There was always great rivalry between neighbours to see who would have the biggest haystack. Later, hay presses driven by engines would create small rectangular bales, and later still farmers would buy machines that would bale the hay into large rolls.

The horses were a big part of town life, too, used for transport and construction. People drove into town in their horsedrawn vehicles to catch the train to Melbourne and to go to the various stores, whose shelves were stocked with provisions delivered by Scottie McPherson. Scottie, a hardworking chap, used to deliver beer barrels to the pubs, lowered into the hotel by ropes through trapdoors in the footpath. He always wore a leather apron right down to his boots and sat at the front of his wagon on an upturned cream tin covered with a chaff bag. Scottie would pick up potatoes from the farms to go to Melbourne by train, carting them to Leongatha where they'd be inspected for grubs and rot before they left.

Then there was the 'cream wagon', pulled by four horses, which plied the district picking up the cream from the dairy farms and taking it back in cans to the butter factory for processing. Everyone had their own big silver cans with their names written on them. The driver would put meat and bread ordered by the families in their cans before he left Leongatha to deliver at the same time as he picked up the cream. In the wet season when parts of the district flooded, the driver would sometimes have to camp on the farms he was visiting until the waters receded. The covered horsedrawn vehicle had been replaced by a truck by the time Merv and Rob were born but the truck was still known as the 'cream wagon'.

Another familiar sight was that of Joe Ryan, the World War I veteran who worked on the roads. Because he worked miles out of town, Joe used to camp where he toiled, patching parts of the road with blue metal, crushed rock or gravel from a tip dray pulled by a team of Clydesdales. Joe was a keen gardener and would plant vegetables and flowers, including red hot pokers, where he camped so that months afterwards you could tell where he'd been by the flashes of red flowers by the side of the road. Joe had fought in the cavalry in the war and gave Rob his bit and stirrups from that time. Rob was about fourteen then and regrets that he didn't ask Joe about his war experiences. All he remembers him saying was that when the horses got shot they just kept going and that when they went down you knew they were gone. Joe was a lovely man, Rob says. He would've cared for those horses.

Horses were part of the shared history of Rob's family life. There was one story, though, that Rob's father always told in confidence—to keep it secret from his wife. Rob's mother was dead against drinking alcohol; her family had been Racobites, part of the temperance movement in Scotland. His father's grandmother had come out from Scotland before the turn of the twentieth century and lived with her husband in Mirboo North, hilly, forested country not far from Leongatha. Her husband would send her into town on the wagon to go to the stores but she'd get on the grog and be so full that she couldn't drive back. The horse would unerringly take her home on the bush tracks and when it arrived with its drunken passenger Rob's great-grandfather would unyoke it, throw his wife over his shoulder and carry her inside.

The era when the dependable Clydesdales were essential wound down throughout the 40s yet Merv and Rob recall that even when the Ferguson tractors became popular, families still kept the horses. 'They couldn't bear to sell those Clydesdales so they stood around in the paddocks for fifteen years until they died. It was a bit sad, really,' says Rob. In places like the Mallee, though, where large numbers were used to plough the wheat fields, the farmers took their Clydesdales to the back paddock and shot them. A wheat-belt farmer might have had 50 horses.

'One fella round here told me he was glad everyone had started to get cars and trucks because the heavy horses had been made to work so hard,' he says. Rob recalls coming back from buying some seed with his father once when they saw a farmer working a pair of Clydesdales that had red raw shoulders. The harness had rubbed the horses' coats off down to the flesh but the man kept whipping them. Kinder owners would cut holes in a harness if part of it was rubbing a sore on the horse's shoulder. 'Dad was really upset to see that,' Rob says. 'He used to work his horses hard but he looked after them, nothing flash, maybe he'd chuck a bucket of water over them and rub them down with a bag.'

Merv's father apparently missed the Clydesdales, though—years later when he retired he bought a three-year-old purebred gelding and started showing it. Merv, then 35, got the bug too and bought a six-month-old colt, and Merv's brother joined in. Father and sons would travel to agricultural shows from Yarram to Dandenong. 'All the main ones,' Merv says. 'Won lots of sashes and trophies, too.' He started hiring his horses out for functions such as weddings and end-of-year parties for anyone from kindergartens to footy clubs.

Merv's wife Kiersten is a fan of the breed, too. In fact, the couple met when Kiersten was admiring one of his horses one day. She'd formed an appreciation of Clydesdales years earlier after going on a gypsy wagon ride for five days. Kiersten, her ex-husband and young children were given the horses and wagon, a few instructions and a map and headed off—with no real knowledge of harness horses beyond the time when Kiersten had ridden Clydesdales as a child, jumping over the fence of the Marriott market gardens in Keysborough to ride them bareback. The owner of the gypsy wagon checked up on them along the way and brought them feed for the horses and supplies. All went well until the fifth day when they were headed home and a storm hit. Kiersten had never experienced anything like it: galeforce winds buffeted the wagon, driving rain lashed at them and branches of trees fell on the road ahead of and behind them. But the two Clydesdales put their heads down and marched on. The younger and less experienced of the pair looked nervous but the older one kept him steady. The two horses made their way home, right up to the driveway, without event. Kiersten, whose hands were rigid with cold, found the owner calmly eating his dinner when they arrived. 'I knew you'd be right, there was no point me going out in it,' he said, certain the horses would bring their passengers home safely.

'They're so faithful and gentle,' Kiersten says. 'They're not like normal horses.'

Merv, too, sings the praises of the horses so often called 'gentle giants'. They're placid, easy to handle and good-natured, he says. They're not cheap to shoe, though, at 200 dollars per horse per shoeing, and mean hard work for farriers who have to support their hooves on stands because they're so heavy to hold. And they do have their foibles, too. One old Clydesdale Kiersten knew of had a terror of camels and would start dancing on the spot if he ever saw one. (There were often camels and camel rides at shows so this was more of a problem than it would seem.) And Clydesdales like eating trees, she says.

Merv owns two purebred Clydesdales now: Milo, who's seven, and Saxon, the son of a stallion, Sarge, that Merv owned for eighteen years until Sarge died. Saxon was a 'surprise' foal who appeared after the stallion's death. But Merv has owned many other heavy horses, one of them the last asparagus-picking horse from Koo Wee Rup, Australia's major asparagus-growing region. George was the only horse left pulling an asparagus wagon but, like the horses before him, he was pensioned off. He came to the Allen home in semi-retirement at twenty years of age and performed light duties for four years before he died. George's death marked the end of an era; no horses are used in the asparagus trade now.

As well as taking them to events and functions, Merv still works his horses on the farm, harrowing pasture and feeding the cattle using the sled, keeping alive the traditions he knew as a child. It may not be practical, it may be a whimsy but it's his way of holding on to the heavy horses.

28

THRILL AND SPILLS OF HARNESS

Neville Gilpin was born with horses in his blood.

Gilp, as he's known, grew up on a beef, sheep and dairy farm in Mirboo East, in the Strzelecki Ranges. Before he was born his father Tom had worked in the heavily timbered country at Gunyah to the southeast, logging with heavy horses on inclines too steep for tractors. One bloke would stand at the top of the hill and attach the felled log to the horse's traces, Gilp explains. Another bloke would stand at the bottom waiting for the horse to drag the log down to be stacked and taken away. The horse would go up and down the hill by itself. But sometimes a log would start skidding behind it, gain momentum and the horse, not wanting to be hit in the back of the legs, would take off in a mad gallop down the hill in front of it. The log would hit the dirt at the bottom and stop, but unless the horse had learned to pull up quickly it would be jerked back in the collar.

Tom broke in horses too to supplement the farm income and support his wife and six children. One of his clients, a milkman in Morwell, enlisted him to train cart horses for his rounds. It would've been in the late 1950s, Gilp says, when his father broke in one of them and handed it over to the milkman, reckoning it was a terrific horse that would pull anything. But one morning within its first few weeks of work the horse side-swiped a parked car with the cart and instead of stopping as it had been trained to do surged ahead, leaving the milkman, who was delivering the bottles, watching horrified as it collided with another five cars. The milko apparently got into a bit of trouble over that.

Tom Gilpin broke in all his children's horses. He travelled to Omeo in the high country once, bought six brumbies and broke in one a year for each of the children. Among that load of brumbies was a blue roan mare who produced a string of very pretty foals. Gilp now has the grandson of that mare. The coat of a blue roan horse is a mix of white and pigmented hair with black underneath, giving it a smokey appearance, with black mane, tail and socks. 'When I was growing up there were blue roans everywhere—I never thought they were special. Now there's not many about.'

His father bought other horses from the saleyards. He told Gilp he once went to a sale where some of the horses were penned in a yard that was missing the bottom rail of the fence. During the sales Tom noticed one of the heavy horses drop to its knees, then on its side, then wriggle under the fence to get out. He thought, 'Anything that smart has got to be useful—I'll buy it!' Working horses had to have some nous about them to know what to do and when to do it without being asked and Tom was right about that one; the escape artist turned out to be a good horse. The family used him for some years ploughing and working around the property.

Tom, who born in 1927, was quite a rider in his time. In the early 1950s he was invited to train with Australia's Olympic dressage team. He used to instruct at pony club, showing the children what to do rather than telling them, and someone associated with the Olympic team noticed him riding at the club one day. But he had cows to milk and kids to look after so he couldn't join the team.

He taught all his children to ride and kept instructing them as they grew up. Gilp remembers breaking in a couple of horses with him once and appearing to fall short of his father's standards. He had saddled one of the horses, which used to buck, in a Western saddle with a typically big horn at the front. He was wearing a baggy jumper at the time. The horse started pigrooting, bucking and carrying on as soon as he got on it, and his jumper got caught on the horn as he was thrown around in the saddle, bending him forward. As Gilp tried to get the horse under control and free his jumper, he could hear his father bellowing at him. 'Sit up straight! I taught you how to ride better than that!'

When Gilp was in his early twenties he went chasing brumbies with some other riders up in the Barmah forest, on the Murray River near Echuca. He and the other riders had found a fair sort of mob, he says, and were pursuing them for a while. Then Gilp lost the mob while he was tailing a horse that had become separated from it. One of the other riders found him and asked him to hold on to a foal he'd roped. Half an hour later a yearling appeared, looking for the other horses. 'I thought, bugger the foal, I'll try to get that yearling.'

Forty-five minutes later Gilp was still chasing the yearling flat-out until his own horse was so tired it could hardly raise a canter. He tried to rope the brumby whenever he got close but kept missing. They finally came to a bend in a creek, cornering the brumby, which Gilp was then able to rope. He thought he had him until the brumby 'found a whole new lease on life' and took off at a furious pace again. Gilp gripped on to the rope as the brumby passed in front of his horse, his horse dropped its head and Gilp flew over it onto the ground. The brumby dragged him along the ground, through the scrub, over dirt, bark and sticks, for 50 metres, with Gilp clinging to the rope until his shoulder lodged against a stump and stopped him being dragged. 'Got you now,' he thought.

But the rope started slipping through his hands until the brumby jumped over a log and yanked it through to the knot at the end. Gilp hung on and the brumby finally pulled up. By now his hands were bleeding from the rope burns, but he had his horse.

He broke the brumby to harness later and it turned on another performance, only this time they had an audience.

Gilp had got the horse used to the harness and had him dragging a sled around successfully. Then came the day to try him out in a cart. Gilp had half a dozen people with him to help and, in the rush to start while they were all there, forgot to attach the kicking strap to the back of the harness. The kicking strap goes from one shaft to the other, over the rump of the horse, to stop it lifting its back legs too far in the air, teaching it that it can't buck in harness. Gilp hopped into the jinker ready to take off but the horse was slow to get going. He leaned forward to give it a smack on the rump with the reins and it refused to move. He did it again and the horse suddenly gave a big heave and took off. Gilp fell back in the seat then forward out of it as the horse stopped, landing face-down, one hand on the horse's rump, the other on the shaft of the jinker. He was still looking down at its rear legs—minus the kick strap—when the horse took off again. Unable to right himself he bumped along head-down, dreading the horse booting him off its rump yet unable to get off the cart because he knew he'd get hit by the jinker wheels. Stuck there for what seemed like way longer than it was, Gilp was eventually flicked off onto the ground when horse and jinker lurched to one side. He gave up the display of harness breaking for the day and put the horse in harrows for a while until he thought he was ready to try the jinker again.

For the past five years or so Gilp has worked several Clydesdales for a friend. He teams up with Merv Allen (see previous chapter) to take a wagon to functions and goes to working horse exhibitions where they give demonstrations of chaff cutting, ploughing, harrowing and stack grabbing. It's hard to find work for big horses now, he says, but he makes as many opportunities to do it as he can, such as riding the Clydesdales the 10 kilometres to town and back, which also helps keep their hooves in shape. He's had a great time with the Clydies, he says, and all without incident!

29

TROOPER: A HORSE GIVEN A FIGHTING CHANCE

Wayne Armstrong, farrier and carriage enthusiast, didn't have to think too hard when he was asked to take part in a program rescuing and training Walers, the breed revered as Australia's war horses. Wayne's grandfather-in-law fought in the Australian cavalry, 4th Light Horse Regiment in World War I and the elderly man had spoken at times of his regard for the horses. When Wayne was told that some descendants of the same stock were being saved from slaughter and trained to help bring their legacy to light, he was in.

The Mitavite Equitana Waler Legacy program took six Walers captured in the outback and allocated them to horse trainers chosen for their skills in starting young and unhandled horses. The horses were among a truckload caught on a cattle station southwest of Alice Springs and transported to saleyards several hundred kilometres away. Horses bought at those sales often went on to the abattoir in Peterborough, South Australia, destined for the dinner tables of Europe and Japan. The six horses that were spared were to be educated in various disciplines and appear at the Equitana horse expo in Sydney in November 2013, not much more than a year after they were caught.

The program came about after Sandi Simons, a riding coach and consultant to Equitana, noticed Walers at clinics she held and wondered why the versatile mounts weren't more recognised as riding horses. She thought that Equitana, the annual equine exhibition attracting tens of thousands of people, would be a fine place to showcase them. The Waler Horse Society of Australia, which champions the cause of the horses, was only too happy to come onboard.

Wayne first heard about horses in the Australian infantry when he was a boy talking to the grandfather of friend—and future wife—Rhonda. Leslie Oliver Langtip, or 'Gang' as Rhonda called him, would speak about his time in the Middle East but never of the horrors of the bombing, shooting or killing, keeping these stories to himself. He did say though, 'If you weren't scared you were an idiot.' Nor did he speak of his gallantry. Leslie was the epitome of a gentleman: quiet, unassuming and always there if you needed him. Of Chinese/Mongolian heritage, he stood 6 feet tall and always made sure that he was physically fit, despite being wounded in the knee during the war.

Leslie served in a 'half-troop' with his three brothers—one of whom kept a diary—in B Squadron of the 4th Light Horse Regiment. When he signed up as a nineteen-year-old in early 1916, he was called into the showgrounds in Melbourne, along with the others in his unit, to be tested to see whether they could ride. The English major who acted as instructor set up a 44-gallon drum on its side and asked the men to mount and dismount it. One of the recruits stepped forward and spat on the ground in front of him, disgusted. 'We ride fine; what about you get on the drum,' he said.

It must have seemed ridiculous to Leslie, too, who'd come from an illustrious horse-riding family. He had won the Australian National Racing Pony championship in the days when ponies were raced like thoroughbreds, and his sister held the Australian high-jump record.

The four Langtip brothers served in Egypt and Palestine, at one time under A.B. 'Banjo' Paterson, who commanded the Australian Remount Squad. Leslie recalled the bard, who was in charge of breaking in horses there, as being a good horseman, a bit of a wag who liked to joke, but somewhat standoffish. Celebrity superiors aside, Leslie spoke of the everyday rigours of war and the way the soldiers would groom the horses on the stomach to try to alleviate their hunger when food was scarce and how they'd share their last drops of rationed water with them.

He was part of the famous charge on Beersheba on 31 October 1917, which later allowed the British Imperial Forces to advance into Palestine. He talked of galloping over trenches filled with barbed wire, bayonets locked into position ready to take on the enemy. He was wounded in the leg not long after Beersheba and taken off the battleground, strapped to a little donkey on a trip taking three days and nights. He remembered it as being a 'kind donkey' and feeling guilty with every step it took because it was so small.

'If you ever get a chance to do a favour for a donkey, luv, do it,' he once told Rhonda. She did, years later, buying an unwanted donkey from a woman who'd lost her property and had no home for it. 'It was such a bad donkey!' Rhonda says, laughing. The woman had been feeding it cigarette butts, for some unknown reason, and perhaps it was going through withdrawals, Rhonda suggests.

Her grandfather rode three horses and one camel in the war; the first two horses died from under him, the third he had shot when the soldiers departed Egypt. The Australians couldn't bear the thought of leaving the horses to the 'Gippos', who they'd seen treat horses cruelly, so put them down instead. It was the worst part of the war, Leslie said. 'You didn't shoot your own horse but someone else's,' says Rhonda. 'Grown men were sobbing their hearts out.'

Sergeant Leslie Oliver Langtip was awarded a DCM (Distinguished Conduct Medal) for his bravery in capturing a Turkish field gun in an advance at Kaukab Ridge in the Sinai Desert on 31 September 1918 and for driving the enemy back with it—or as the original draft of the citation said less delicately, shooting the Turkish party with it.

After the war, Leslie changed his surname to Langton and settled down to family life but he retained his love of horses. When Wayne and Rhonda married in 1974 and set up their trotting stables in Bacchus Marsh, he would travel from Malvern, a suburb of Melbourne, to visit them and loved to sit and listen to a stable full of horses munching away on their feed. It was during these visits that he would talk of the Waler horses he rode during the war. He last rode when he was 80, after a break of 30 years or so, mounting one of Rhonda's dressage horses, an 18-hand giant. 'It was just like he'd slipped back into an armchair,' she says.

Leslie was an intelligent man and somewhat of a genius with numbers. He looked after the books for Rhonda's father's businesses from his 60s to when he died at the age of 81 in 1977, and could run his fingers down a column of ledger book numbers adding the tally up as he went. He died at his desk, ledger book open, with a pen in his hand. His heart had simply stopped.

His grandfather-in-law's Anzac history was one reason Wayne Armstrong signed up for the Waler program. He had also become aware of the plight of the horses. 'I thought it was wonderful to be asked and it was a wonderful way to save horses that were getting shot or were left to die of starvation,' he says.

His own experience with horses—45 years of it—made him an ideal candidate. The children of a trotting trainer, Wayne, his four sisters and two brothers started driving sulkies as soon as they could reach the footrest to climb aboard. Before then they sat on their father's knee and drove. Wayne trained his first horse to harness as a fourteen-year-old after his father had a heart attack and Wayne stepped in to help him, and drove his first winner at sixteen. He opened his own stables at nineteen with Rhonda, an accomplished rider who had successfully competed in a variety of horse sports, winning at the Royal Melbourne Show, and showjumping with the likes of Eric Musgrove and Ern Barker. Leslie Langton would always listen to the harness races on the radio when the couple's horses ran. After his death they found a scrapbook full of newspaper cuttings of their race wins or mishaps that happened. Along with the news clips were tape recordings of races that Wayne drove in—whether he won or not! They also found the DCM medal for valour and gallantry and its citation buried among the belongings.

The stables went well for the first few years until the Armstrongs' star horse broke a leg and another successful horse broke a bone in its foot. By then the couple had tired of working thirteen-and-a-half day fortnights. Wayne switched to farriery in 1978.

He kept working with horses in harness, though, through his passion for the sport of 'combined driving', working his way up from a single horse to competing with a four-in-hand. Combined driving, championed by Prince Philip, is similar to eventing, Wayne explains. Drivers compete over three days in dressage, cross-country marathon and obstacle driving. Wayne was the state and national title-holder at the time of writing. He also teaches mouthing and long-rein driving, and has taught classes at Equitana in this, and instructs on handling. 'We're given a lot of horses people can't handle and re-educate them,' he says.

Wayne knew little about the horse he was to educate in the Mitavite Equitana Waler program beyond that it was unhandled, buckskin, a colt and that it already had a name: Trooper.

Those who captured him thought Trooper was about three years old, judging by the amount of scarring on his body. He arrived at Wayne and Rhonda's property in Bambra on 7 November 2012, thin, ribs showing, and covered in scars: bite, kick and scratch marks, with a bad knock to his nose and a torn ear. Wayne, who likes buckskins, was pleased to see that the colt had a short back (good in a carriage horse), and was a compact, nuggetty type, which he also prefers.

But Trooper came with a warning. 'Be careful of that horse, he's dangerous to people and dogs,' said the man unloading him, who refused to get into the back of the truck to guide him out.

Wayne wasn't daunted. 'You'd expect a bit of trouble from any horse taken from that environment and from a herd.'

He soon discovered that rather than exploding or running away if frightened, Trooper would stand and snort, conserving energy, as he would do in the desert. Even when he was released into the paddock with the Armstrongs' thirteen other horses, he never galloped after them or bucked but walked everywhere. Still does. He's not fussed about what he eats either, heading for the nearest clump of grass rather than moving around searching for the tastiest blades.

Wayne put Trooper in a small yard when he arrived, with a pony for company. He handled him for an hour and the colt didn't once attempt to bite, kick or strike at him, which he had expected him to do. He stood on the opposite side of the fence at first, in case Trooper reacted in the wrong way, slipping a rope with a ring around it over his neck so he could get used to feeling it tighten and loosen. He replaced the lasso with a rope halter and put a breaching strap around his hindquarters—Trooper's first lesson in leading. 'Within five minutes he was walking beside me like an old horse.'

He stroked Trooper with a towel attached to some poly pipe, desensitising him. Trooper pulled back a couple of times then stood quietly.

The next day Wayne strapped Trooper's front leg up—a technique that puts the horse on three legs and stops the flight response—then handled him all over his body. He strapped a back leg up then guided him to the ground, where Trooper lay as Wayne patted and rubbed him all over and sat on him; more desensitising. 'Trooper is the first horse that I've ever put leg straps on and he didn't do anything—usually a horse will kick out,' says Wayne, pointing out that everything is done gently.

He groomed Trooper and picked up his feet on the third day and had him gelded at the end of the week. Wayne had to go into hospital at the time and asked a friend to help the vet, who charged extra because Trooper was a wild horse. Trooper reacted when the vet put the first injection in his neck, rearing up, pulling back and kicking a pole behind him. 'I would have put a leg strap on,' says Wayne.

After a week's handling, Trooper would lead, tie up, was comfortable with being groomed all over and having his feet picked up. Wayne rested him for two weeks then resumed training, teaching him to lunge, introducing mouthing gear then long-reining him in this.

During Trooper's fourth week (his second week of training), Wayne introduced a roller, crupper and bridle, and started long-reining, driving him around the yard, teaching him to flex and turn, and stop. By the tenth day of training he was taking Trooper out on the roads, driving him from behind for 3 or 4 kilometres. He got him to drag a tyre to get used to the noise and to teach him to pull, put him in blinkers then long-reined him through steep gullies and across creeks. Other horses would have baulked at this, says Wayne, but Trooper just went straight through them. He's a quick learner, picking up commands and expectations, such as reining back and standing still at a gate while Wayne stops to close it.

In week six, Wayne got Trooper used to a long-shafted steel cart, tying him to another horse in the shafts, and struck trouble; Trooper didn't like the steel shafts touching his sides and kicked. It took him a couple of days to settle and to stop lashing out before Wayne was driving him around the yard, sitting in the carriage. He ventured out of the yard after three days, taking Trooper for 8-kilometre drives through the state forest, up and down hills and out on the roads. 'He was a star,' Wayne says.

Wayne was equally amazed at the horse's progress when he started riding him. 'He accepted everything—I couldn't believe how good he was.'

The trainer spelled him for six weeks to let him 'think about things', while he prepared his team of Welsh mountain ponies for the National Cross Country Carriage Driving Championships. When he resumed training he started riding Trooper over some poles then jumps—and found he took well to jumping, too.

'He's very good,' he says. 'A very relaxed customer. The breed has to play some part in the way he accepts things like he does.'

Wayne doesn't believe in giving horses treats nor does he expect a horse to bond with him or become his 'best friend' but he says Trooper does come when he whistles, though curiously only when he's got his halter on.

Wayne isn't sure yet what he will demonstrate at Equitana with his laidback little horse but has no doubts about him performing there. The Walers will also be exhibited in a 'breed' pavilion so that visitors to the expo can see them up close. Wayne, of course, is already taken by the breed.

'I'd be willing to take on more Walers,' he says.

Apart from the benefits to the breed, Trooper's training has brought alive memories of the man the Armstrongs still fondly think and talk of as Gang, 35 years after his death. Wayne likes to think that Leslie Langtip is looking down and smiling at him because he took on the challenge of working with a wild Waler, possibly a relative of the horse Leslie rode during the charge of Beersheba, and has given it a second chance.

The Waler's tale

The Waler breed had its origins in horses brought out to Australia on ships in the First Fleet in 1788, collected from the Cape of Good Hope. These horses were later bred with horses including thoroughbreds, Timor ponies, Arabians and heavy horses such as Clydesdales and Percherons. The result was a uniquely Australian colonial horse, a 'type' that became known as the New South Walers or 'Walers'. Successive generations defined the Walers as a 'breed', that is, when progeny breed true to a certain type and meet a standard that is set down.

Tens of thousands of Walers were bred to be used for transport in the developing colony and to be sold as remounts for the British cavalry, with strict standards set for temperament and conformation. The result was a robust, sensible horse that could handle long journeys at pace, and which was well-suited to desert warfare because it could live for long periods with little food or water.

The Australian Army enlisted the Walers for its cavalry in the Boer War, and in World War I when they became famous in the Australian Light Horse regiments for their feats of endurance and courage, most notably the triumphant 3-kilometre charge across the searing deserts on Beersheba in 1917.

Even the most infamous Waler in the war—Bill the Bastard—distinguished himself in such a feat. Bill, the horse ridden by Major Mick Shanahan, was a horror. He could appear to be dozing yet boot anyone who approached within striking distance of his back legs, and always bucked when asked to gallop. But Bill redeemed himself during a battle at Romani, Turkey. The major had found four Australian soldiers out-flanked by the enemy, their horses shot or lost, and needed to get them to safety quickly. Bill carried all five men—three on his back and one on each stirrup—over a kilometre through soft sand, galloping laboriously out of the firing line. One of the horses in the Mitavite Equitana Waler program has been called Bill as a tribute to this brave horse, with other horses in it given names of famous Walers from the past.

About 160 000 Australian horses, three-quarters of them Walers, were sent overseas during the war; only one Waler came back. Due to quarantine restrictions, only Sandy, the mount of a major-general who died at Gallipoli, returned, shipped back to Australia in 1918. The horses were often shot by soldiers who couldn't bear the thought of leaving them behind to uncertain fates.

Despite their noble past—Walers were once held to be the best cavalry horses in the world—they were all but abandoned as a breed in Australia by the 1960s. They made good stock horses and were used by the mounted police but were overlooked as pleasure or sport horses in favour of thoroughbreds and imported breeds such as quarter horses, European warmbloods and Arabians.

But interest in the breed revived in 1971 when the Australian Stock Horse Society formed and put Walers on its books, and in 1986 with the formation of the Waler Horse Society of Australia (WHSA).

The WHSA founders had heard that Walers on stations in the Northern Territory were being slaughtered and appealed to the station owners to be allowed to take them out. The Walers had been reduced to being regarded as feral pests, yet these horses were the direct descendants of the stock that once bred remounts, genetically isolated since 1945. A stud book was established that lists horses and their progeny derived from the old bloodlines of these remnant herds.

The horses retained the attributes that Walers were once lauded for: good conformation, temperament, strong bones, frugality, intelligence, courage, providing a comfortable ride and versatility.

Elizabeth Jennings, Waler enthusiast and breeder and WHSA president, says there are variations within the breed now, like there were during the wars when individual horses were categorised according to their use: Waler ponies for scouts, light Walers for officers, medium ones for troupers, and heavy Walers as artillery horses. Today, they make fine all-rounders for anything from polocrosse to eventing, to dressage and carriage horses, then a trail ride on the weekend. Elizabeth says Walers are good companion horses.

'That's why so many soldiers were heartbroken when they had to leave their horses behind. They'd talk about them like they were people. They were mates on the battlefield. If I'm upset I'll go out in the paddock and my horse knows—the relationship the soldiers would have had with their mounts would've been the same.'

PART IX

Beating the odds

30

BRAVE ANNE MEETS HER MATCH

When Anne Skinner took up the job as a state coach with the Riding for the Disabled Association in the late 80s, it seemed like a perfect match. The RDA was expanding and needed an extra coach in Victoria to work with its volunteers and riders; Anne had been working with people with disabilities, had been a riding coach and loved horses. She started the new job with characteristic gusto.

On any day of the week she could be found travelling to Dimboola in the state's west or Orbost to the east, on the road visiting RDA centres throughout the state, working hands-on with the riders, their carers and the volunteers who ran the sessions. Later she became a national assessor of coaches, visiting RDA groups around Australia. She'd watch, amazed, as children who'd never talked before spoke their first words, or as they learned to count when the horse they were riding stepped over 'one-two-three-FOUR!' poles on the ground. She saw how riders who had started slumped in the saddle sat upright in the space of weeks. She would enthusiastically tell their parents about how the motion of a horse alone could stimulate and strengthen muscles and improve coordination and balance, and more. She told them these things from rote, as she herself had been told. But it wasn't until she became disabled that Anne Skinner really understood them.

Anne was born with a love of horses, something her mother said—and hoped—she'd grow out of. 'She went to her grave thinking that!' As soon as Anne could read she was circling ads for horses in the For Sale columns of Sydney's *Telegraph* newspaper. The horses were advertised for about 30 pounds each but beyond that Anne had little idea what the ads meant—owning a horse was just a dream. An incident on her cousins' farm when she was about nine firmed her resolve to have a horse of her own. The cousins had a pony called Thruppence but weren't interested in riding her themselves. Anne took her chance one day when she was out picking mushrooms with her father.

'You go on, Dad, I'll close the gates,' she offered. 'I'll catch up with you.'

As her father drove off, Anne ran back to the sheep pens where Thruppence was yarded, hastily made up a bridle from some rope she found, led the pony out and scrambled on. Thruppence promptly took off, pelting across to the other side of the paddock in the opposite direction to Anne's father, heading for the tree with the lowest branches. Anne clung on, not wanting to yell out, and slipped off when Thruppence came to a halt then casually walked the pony back to the yards. Cool as a cucumber. Her father said nothing at the time but years later laughed as he recalled the sight of his small daughter streaking across the paddock. The experience with the bolting horse didn't deter Anne, as gutsy then as she is now. 'I thought, "Nothing will stop me now."' It was a vow she was to repeat at times throughout the rest of her life.

Anne bought her first horse at seventeen after learning to ride at Cannings Riding School in Mildura; she had lessons in return for working, taking out trail rides and caring for the horses. She'd been told about a horse that was for sale, a buckskin and white gelding that cost 100 dollars. That was all she needed to know. The horse's name was Terror—although she called him 'Terry' in front of her parents—and was 23 years old. But being aged didn't seem to bother him. Terror, by name and by nature, was still going strong at 33.

Anne competed on him in her late teenage years, but had to set up a pony club to do it. When she was eighteen she and a couple of friends established the Alcheringa Pony Club over the river from Mildura at Gol Gol. 'I had an ulterior motive,' she says. 'I wanted to compete at pony club level for a few years while I was young enough.' The club went on trail rides, camps, held game days and was generally very social, she says. But the opportunities to compete outside it were limited—Adelaide was a five-hour drive and Melbourne seven hours away—leaving the annual shows at Mildura and Wentworth and a gymkhana at Pooncarry, a bush town, as options. Anne entered mainly hacking and games events—Terror took to novelty races, charging off from a standing start at the word 'Go!' There was no dressage. 'I knew dressage existed but very little about it,' she says. She stayed on as a coach at Alcheringa, which is still running today.

Anne married Rex, a policeman, when she was 21, converting him to riding. Rex, who played cricket and football, had no choice. 'If you didn't like horses, you had no chance with Anne,' he says. Anne recalls taking him riding before they were married, borrowing a horse for herself and putting him on Terror. The couple were enjoying a rest, sitting on the horses under the shady gums of the Murray River as the horses munched, when Anne was telling Rex a story that involved the word 'go'. Knowing what Terror would do when he heard it, she spelt 'G-O' but—*whoosh!*—Terror shot off anyway, causing Rex an 'undignified dismount', as she puts it.

The couple moved to Melbourne for Rex's work soon after they married and had two sons, Anthony then Tim. Anthony was horse-mad from the get-go, says Anne. She recalls him coming up to her one afternoon when he was four and she was pregnant with Tim, lugging the saddle for his pony Minihaha, stirrups dragging behind him, remonstrating that they hadn't been for a ride that afternoon. Anne was then working full-time as a 'house mother' for the EW Tipping Foundation, a foundation that provides residential homes for adults with physical and intellectual disabilities. She took up the RDA job when her sons were older and it was becoming harder to juggle running a residential home and taking them to sport and other activities.

She decided to get serious about showing and bought a stock horse foal and trained him from scratch. Papermate eventually won champion stallion at the Melbourne Royal Show and led out the stallion parade. In the mid-90s she turned her attention to dressage and three-day events, competing on a handsome dapple-grey Percheron-cross called Ted that Rex had given her. Rex had raised Ted from a foal but was endurance riding and needed a lighter mount; Anne had a rapport with the horse so it seemed a better fit. By now she was working for the RDA and used Ted there, too.

But in 1997, Anne and her family's world changed forever.

It was a Friday evening, 19 September, when the accident happened. Anne had been delivering goods to RDA centres around Melbourne, towing them in her horse float, and had just arrived back at the family home in Drouin in Gippsland. She was due to compete in a one-day event on Ted the next day at Werribee, and was thinking about all the things she had to do beforehand. Rex was working in Melbourne at the time. Anne backed up the float beside her stable, ready to load Ted early the next morning, manoeuvring it to exactly where she wanted it when her mobile phone rang. It was Rex, saying he'd just finished work and would Anne like to meet him for dinner on the way home. She replied that she'd just got home and had to pack the car for the show the next day so it was no to dinner and she'd see him soon.

Distracted by the call, she got out of the car, leaving the motor running with the engine still in reverse, walked to the back of it and wound up the dolly to disconnect the float from the car. As soon as the car was free it started moving backwards, nudging the float out of the way, knocking Anne over and taking her with it. It rolled her for 7 or 8 metres until it stopped, lodged against her body. Anne was caught under the car's rear axle, her face forced into her left knee as the exhaust burned the back of her head, her right leg at right angles to her body. One and a half tonnes of station wagon still trying to move. Anne's first thought was that the car had hit her but she had no idea how it had happened. Then, 'Oh, gee, I wish they'd get this car off me. I'm going to be awfully stiff on the cross-country course on Sunday.'

Dusk fell and she was still pinned under the car, falling in and out of consciousness. Breathing was difficult but she was not aware of any great pain. She focused on the headlights of moving cars on the road at the end of her driveway, willing them to turn into the drive before each set disappeared.

'The next one will be Rex. He'll drive in the drive. The next one will be Rex,' she said to herself, wondering how long it would be before the full tank of petrol she had put in earlier would run out.

An hour-and-a-half passed before her husband came home.

Rex recalls driving in and seeing the float light up in the headlights. Nothing seemed out of place. He went inside and asked Tim, who had come home a few minutes before him, if he could go out and find his mother, give her a hand, and ask her if she wanted him to put the dinner on. Tim, then seventeen, found Anne under the car and ran back inside, yelling. Rex rang an ambulance, burst through the closed back door and ran out to his wife, crouching down and reassuring her that they'd get her out soon. Anne started to go into respiratory failure.

Rex knew that if he left the car there his wife would probably die; if he didn't her fate could be the same. He decided to act. He found a jack and jacked the car up one notch at a time, easing its frame off Anne's body, instructing Tim to push heavy drums under the corners to stop the car from rolling. Anne can't recall being in severe pain up until then but when the car was removed it was excruciating.

The paramedics stabilised her while arrangements were made to transport her firstly to the local West Gippsland Hospital and from there by helicopter to the Alfred Hospital in Melbourne. She was diagnosed as having a lacerated liver, broken ribs, a suspected fracture of the neck, crushed lumbar spine, crushed sacrum and fractured pelvis. Full of drugs, Anne's only recollection of that time was being wheeled into an elevator and looking up at the bright lights and people carrying clipboards thinking she was on *Sale of the Century*.

Anne was in and out of theatre every couple of hours for surgery over the next 30 hours. Rex, wanting to be close, smuggled a swag into the hospital and camped in a storeroom. On the Sunday night, after Anne had been on life support for 48 hours, he was advised to call Anne's family in to say goodbye to her. The doctor in charge told him they had done all they could and that his wife wasn't expected to live. Not only were so many of Anne's bones shattered but there was massive internal bleeding and the doctors couldn't find the source of it.

'You do your job and she'll do hers,' said Rex, in the first of several disagreements he was to have with the doctors and staff. 'Give her a chance, she'll fight. I know—we've been married for 27 years, I've got the bruises to prove it. She's a tough girl.'

Anne hung on.

She was put in a drug-induced coma for seven weeks. When she was slowly brought out of the coma and eventually regained consciousness, she realised she had no feeling from the waist down.

Anne's spine had been completely separated from her pelvis and the lumbar spine and crushed into dozens of pieces. Nobody had survived those injuries long enough for a surgeon to even contemplate how to put a spine like hers back together again.

Ten days after she came out of the coma an orthopaedic surgeon, Owen Williamson, came across Anne by chance and took an interest in her case. He'd never before seen the combination or complexity of spinal and pelvic injuries that Anne had. Nor did the equipment exist to deal with them. Undeterred, he took over, enlisting an American company to design special plates and a bolt to join them, and devising an operation that would enable a team to do this. He told Anne that the outcome was uncertain and that he didn't know if she would walk again. As they operated, the doctors discovered that Anne's spinal cord wasn't severed, just damaged. The ground-breaking operation that reconnected her bones made news around Australia. A couple of weeks later when Anne was having her foot massaged, she realised with joy that she could feel her left leg.

But it was barely the beginning of her recovery.

Next came the long, arduous process of learning to walk again. The following months in rehabilitation were the hardest in her life. Progress moved at a glacial pace and sometimes backwards. Even walking two or three steps along the parallel bars, lifting and dragging one leaden foot a few centimetres after another, exhausted her to the point of nauseousness. As the weeks of rehabilitation dragged on, Anne despaired that it would ever work and about her quality of life in the future. She was on a drip of antibiotics for two years and is still on a maintenance dose of morphine to keep the constant pain at bay. She became seriously depressed, not wanting to see anyone or talk to anyone. But the thought of being able to get back on her horse Ted helped her keep going.

Anne rode again in June 1998, nine months after her accident. She was still in considerable pain and physically vulnerable but was determined to do it. But first she had to get her surgeon Owen Williamson to sign a consent form for the centre where she would ride. He said, 'What's this?', and after Anne explained, asked her, 'What if I say no?'

'It won't make any difference,' said Anne.

Rex, too, knew better than to try to stop his strong-willed wife.

The first ride was a 'big production', says Anne. The centre with facilities for riders with disabilities was in Ocean Grove, several hours away. As they drove there Anne wondered whether what she'd learned about—and taught—at RDA would actually work. 'I was terrified that all the things I'd been saying all those years weren't true, that they were exaggerations,' she says. 'What if I'd just been saying those things to people for eight years because that's what I was told to say?'

The program she was about to undertake was a form of physiotherapy called hippotherapy. Anne knew the drill: the rider is wheeled up a ramp in their wheelchair, helped out of it and onto the horse then supported on both sides by a 'side walker', while being treated by either a physiotherapist or occupational therapist. The horse is driven from behind with long reins. But it was only when she got to the top of that ramp that she realised how terrifying the gap between it and the horse could seem. 'In your mind you can easily slip into that gap and once you were there you couldn't support yourself.'

After a lap of the arena Anne was exhausted but exhilarated. For the first time after her accident she felt her body lengthen and feel free. 'Walking around the arena feeling the horse's movement I thought, "Yes, I can do this now. I'm not crumpled".' She dismounted after five minutes and it was like her muscles had somehow straightened. All the reassurances she had once given other people had proved true. 'I knew I was on the right track.'

Rex built Anne a ramp with a platform at the top for her to be able to mount a horse at home. He had given Ted to a para-equestrian rider to look after and to compete on when Anne was in hospital and the only consideration was her survival. Realising the family was in for a long haul, he also arranged for friends to take the dog, cat and horses. But Anne had been given another horse while she was in hospital. 'Vengie', whose real name is Unproven Vengeance, was ridden by David Middleton when he was competing in high-level eventing but at sixteen the horse's health and safety was at a much greater risk. David loved Vengie too much to expose the horse to unnecessary risks. 'So he gave him to Anne Skinner, who was in hospital and wasn't expected to walk again, so he could have a very nice life in the paddock,' Anne says. 'I didn't understand that Vengie was supposed to be retiring to our place so the first thing I wanted to do when I got home was to ride him!'

Rex caught and saddled the horse and took Anne on small walks around the paddock on a lead, which she enjoyed. But the day they decided to take the lead off Anne was riding close to the fence and a truck roared past. Vengie did an almighty shy sideways, Anne stayed on, and he stopped dead, as if he was terrified of what he'd done. She says he stood stock-still, four legs on the ground, heart pounding, as if he was saying, 'For God's sake, she's not right, put her back on the lead again!'

Anne had to learn to ride in a new way to compensate for her crooked body. She found she had no pelvic tilt so could no longer swing her hips to, say, propel a horse into a canter, nor could she do a sitting trot because her spine is fused in one part and can't absorb movement. She couldn't feel what her left leg was doing and balance was an issue.

Worse still, 'My head still thinks my body is straight,' she explains. 'It doesn't understand that my hips aren't where they were or my pelvis doesn't tilt whatsoever.'

Yet the horses she rides understand.

She got Ted back and recalls riding with some other people on the road one day. 'Ted went to trot to keep up with the others but I couldn't do it—he broke into the slowest of canters. It must have taken an enormous amount of strength for him to stay in that canter,' she says. 'It never ceases to amaze me how horses can perceive our needs.'

In November 1998, less than six months after Anne started to ride again, she competed in the RDA State Championships in dressage in Grade Two on both of her horses and came first and second. Para-equestrian dressage is judged in the same way as the able-bodied event with the judge scoring the horse throughout the test and giving the rider a single score out of one hundred at the end of it. Paralympic riders are graded according to a set scale of mobility, strength and function and are classified in one of five grades, Grade One being for the athletes with the most severe disabilities. Grade Two riders are impaired in one or both limbs, top or bottom, and have problems with balance, and perform the dressage test at a walk or trot. The riders are allowed to compete using slightly modified tack. Anne uses two whips, as 'legs' to guide the horse on its flanks.

Anne won the Grade Two State Championships in March the next year.

But despite the wins and the gains she was making, she was still in pain and was still depressed. She could no longer work in the job she loved, and as much as she struggled to regain her independence she was mostly wheelchair- and house-bound. Rex would come home after work and find her curled up in the garden, where she'd been watering the plants. Anne was so determined to keep gardening that she'd climb out of her wheelchair and crawl around, even though she knew she wouldn't be able to get back into it and would have to lie on the lawn until Rex got home.

In mid-1999 she felt flattened by it all again. 'It was hard to even get out of the house. We were coming into winter and my body didn't like the cold any more.'

She needed a goal. Anne set her sights on the Paralympic games.

Mary Longden, who was the paralympics head equestrian coach for Australia in the 1996 games, suggested that Anne try out for the Australian squad for Sydney 2000. Anne renewed her efforts, going to as many dressage competitions as she could, bringing home her score sheets, looking at her marks out of ten for each movement, highlighting anything below seven and comparing the comments of the judges. She trained hard and battled the fatigue and pain that would flare up when she took herself off morphine to compete. Under competition rules she wasn't allowed to have any drugs in her system and would have to stop taking the painkiller a week before an event. Rex says she paid the price for this, being bedridden after events, but that she never stopped pushing herself.

She won in Grade Two dressage at the RDA National Championships in October 1999 and in early 2000 travelled to New Zealand to compete in what was the last qualifying competition in the world for the Sydney Olympics. The squad was named in July that year and Anne Skinner was in it.

'It's still really hard to even believe,' she says. 'I wasn't even disabled at the time of the previous games in Atlanta.'

She says she will never forget the roar of the crowd on 15 September 2000 when the Australian team entered the arena at the opening ceremony. 'It was deafening. I get goosebumps on my neck still thinking about it.' Riding with the team on her motorised scooter, she couldn't see the crowd for tears. Rex, watching his wife in the arena in her green and gold Australian uniform, cried openly.

'Australia's Anne Skinner', as she'd become, finished tenth with her team and seventh in the Grade Two dressage individual championship riding a borrowed horse called Lady Luck. It was almost three years to the day that she'd had her accident and, as terrible as it had been, it had taught her that she could achieve things she never thought possible.

Next came the 2003 FEI World Equestrian Game championships in Belgium, then the Paralympics at the 2004 Athens games, where she was reclassified as a Grade Three competitor after her strength improved. Grade Three riders canter. Anne rode another borrowed horse and came eighth in the individual event and ninth with the team. The games and performing in Europe was exciting but it didn't match up to the Sydney games, she says. 'I don't think anything ever will.'

After Athens, the rules for Paralympic Games changed internationally. Competitors would no longer ride a horse out of a pool of 'borrowed' horses, as had been the case, but would need to find their own horse, either owned, leased or borrowed of a suitable quality to go to the Paralympics. Anne was competing in Australia on Laddie, a beautiful black thoroughbred who looked impressive but didn't like dressage. She needed a more suitable mount. Mary Longden got onto the case. Mary, a straight-talking Brit, summoned Anne and her coach Julia Battams to her property, Longhope Lodge. It was June 2006 and the Victorian State Championships were coming up.

'Julia,' she said, turning to the apprehensive coach. 'What are you going to give Anne to ride at the State Championships because that horse of hers is useless!'

Julia replied that she didn't own any horses suitable for Anne.

'What about the Queensland horse?'

'The Queensland Horse', as he was always called, was an 18.1-hand warmblood, a giant of an animal that had spent six months in the stable at Balmoral Equestrian Centre for sale. Anne trained at

Balmoral and knew how big he was and was completely disinterested. She didn't want a horse that she'd need help to handle or mount—she was too independent. She had walked past the horse's stall many times in those six months without so much as a glance, let alone knowing his name or his history.

She had no idea that Aubaine Cossack was a horse that had been in an accident and, like Anne, was lucky to be alive . . .

Friday evening on 17 October 2003 in southeast Queensland. Vet Julian Willmore answered a call from the Allen family, who lived five minutes down the road to say one of their horses had been hit by a car and could he come urgently because there was 'blood everywhere'. Cossack, who was four at the time, had been in a paddock at the rear of the Allens' house when an older horse he was with pushed open the gate, as he'd taught himself to do, and ran out past the house and onto the road in with Cossack following. It was dark and Cossack, a deep brown colour, was hit by a four-wheel-drive on the road. The family, realising that the horses had got out, hurried down to the road. Moira Allen recounted how Cossack came towards her in a 'friendly, sad way', which was unusual for him because he was an arrogant horse with a lot of attitude. She could see and hear blood gushing out from his face onto the road.

Julian Willmore treated Cossack at the time then referred him to a surgeon at a veterinary hospital to treat a fractured jawbone. Cossack's upper teeth had been pushed down into his mouth and his jaw bone fractured through. Without surgery he would never have been able to eat properly again. The surgeon, Dr Tim Hill, put stainless-steel wires in Cossack's jaw and tensioned them to try and get the bones, upper jaw and teeth aligned again. The surgery cost 20 000 dollars but the fracture healed well after the wire was removed, Cossack's teeth were back in alignment and, apart from a dent where the bullbar of the four-wheel-drive hit his nose, he was the same as he was before the accident. The family later decided to sell him, with the proviso that he went to a good home. He was put on the market and sent down to the Balmoral Equestrian Centre.

Mary Longden insisted that Anne go down to the stables at Balmoral with Julia to look at the horse. Anne recalls Cossack's massive dark head and neck appearing over the stable door as she and Julia walked along towards it. 'I looked at him and said, "He's just too big, I'll never be able to ride this horse".' Mary was due to arrive in half an hour.

Julia helped Anne mount him. Anne gave the horse a loose rein and noticed after almost a lap walking the arena that he 'went into the shoulder-in', or was moving sideways as he walked. She wondered what he was doing but Julia could tell; Anne sits with her right hip over the centre of the saddle slightly because she can't feel her left leg. Cossack felt the weight to one side and thought she wanted him to go sideways. Then, he straightened himself up. Julia was incredulous. Anne sensed it, too. 'It was like he was thinking, "She can't really mean that" and straightened up,' says Anne. Anne pushed him on, and Cossack launched into a big, ground-covering trot before sensing that she didn't want that either and pulled back into a 'granny trot'.

Anne felt a connection with Cossack but was still reluctant to consider buying him because of his size. As she rode the horse she started to think, happily, about all the things that Mary wouldn't like. Cossack has a blaze that veers down one side of his face, which could make him look crooked to the judges when he was coming down the centre line in an arena. He was a bit base-narrow at the front—his front feet were too close together, meaning he'd be prone to 'paddling' or throwing his front feet out to the sides. 'She'll pick on that!' Anne thought. His size might make him appear a bit cramped in a 40-by-20 metre dressage arena. 'Things are looking good!'

Mary arrived, stood at the end of the arena and issued a few commands. 'Turn right, turn left. Now canter.' Then she said, 'He'll do. I have to go.' And left.

Anne borrowed Cossack to ride in the RDA State Championships three weeks later, and won. 'He was wonderful,' Anne says.

In August 2007 she went to Belgium and Holland with Julia to compete in the FEI Para Equestrian European Dressage Championships on another borrowed horse. Europe is the heartland of the warmblood horse and Julia organised for Anne to try out a few mounts that were for sale. Anne rode about eight horses, all priced at more than 100 000 dollars each. But as she tried out horse after horse she found herself saying, 'But it's not like Cossack.'

She returned to Australia in October and drove to Balmoral thinking of the big horse. 'If he doesn't recognise me when I get there I'm not meant to have him,' she said to herself.

'Are you there, Cossack?' she yelled out as she walked down the line of stalls. There was silence, then a big, throaty neigh. Anne cried.

She'd found her horse but she didn't have the 45 000 dollars needed to buy him. 'I believe, though, that if you want something badly enough, it will happen,' she says. It took her seven months to raise the money and a nervous wait as two other parties appeared who were interested in buying him, too. A friend's generosity made it possible.

Anne knew Wynsome Reed from her RDA days when Wynsome was a volunteer. Wynsome had bought Anne her top boots to compete in at Sydney and helped her again when it came to buying Cossack.

Wynsome, who turned 82 in 2012, had become interested in horses again after a long break; she'd last ridden in her early 20s. She got back into riding, read up on dressage and studied DVDs, and was soon riding Cossack.

'Cossack's so gentle with her,' says Anne. 'And she's such an amazing lady.'

Cossack loves being involved in whatever's going on and being ridden, Anne says. 'If he sees a car with a float, it's like, "Can we go somewhere?"' He marches out on a trail ride 'full-bore' and is interested in everything he passes. Only once did Anne see any lasting effects of Cossack's accident on him. They were riding along the road one day at dusk, a little later than they would normally have been out, when Cossack became agitated at the sight of some headlights coming towards them on the road, as if remembering that night. Anne felt him wanting to run sideways and could feel his heart beating wildly from under the girth. 'He was terrified,' she says. She stayed calm and breathed deeply, and he settled. 'He defers to me for his confidence,' she says. 'If I'm confident about something, even if he's worried, it won't be a problem.'

Anne competed on Cossack before the Beijing Olympics in 2008 but there just wasn't enough time for her to get the quality of work done that was required to make the squad and she missed out. She was selected to train with the squad for the London Paralympics in 2012 but when the time came for the final selection in April that year, she tripped over a lead rope in the stable and broke several ribs, ruling her out of contention. Missing out again was disappointing and meant shifting her sights on another goal. She says, too, that while the Olympics are the biggest, most high-profile events in the eyes of the public, the World Equestrian Games, held every two years between the Olympics, are considered as, if not more, prestigious in the equine circles. It's like comparing Olympic tennis to Wimbledon.

In 2012 Anne was reassessed as a Grade Two athlete after her spine deteriorated and the strength and coordination in her left side became much weaker. Although she can walk unaided and without an obvious limp, she relies on her electric scooter to get around her property and takes it to all competitions. Her walking stick is never very far from her side.

Cossack keeps improving under her training. He was graded elementary going into medium when she got him; now he's an advanced dressage horse. (Dressage horses in Australia are graded from preliminary (for beginners), progress to 'novice', then 'elementary', 'medium' and 'advanced', according to points the judges award them for a set of prescribed movements.) Rex says that despite the number of good horses Anne has owned over the years, she's never bonded with a horse the way she has with Cossack.

Anne's current challenge is to compete in able-bodied dressage to gain competition experience as there are very few para-equestrian competitions. She hopes to qualify for the FEI World Equestrian Games in Normandy in 2014 and the Paralympics in Rio de Janeiro in 2016.

To Anne it's vital to have goals but she says she's not competitive at all costs and since her accident tries to enjoy every minute of whatever she does. Competing or not competing, 'it doesn't make any difference to Cossack and me'.

31

CLANCY

'Slow down, Mum. There's something wrong with that foal.'

It was a Sunday afternoon in mid-August 2009 and the Crombie family were heading home to Metung from their holiday cottage in the Bundarrah Valley in Victoria's high country. They were passing through their neighbour Helen Packer's property when fifteen-year-old Maddie saw the dark form of the foal lying in the paddock below the road. As the other horses on the riverflat huddled with their backs to the driving rain and wind, the foal lay listlessly on the ground, trying feebly to raise its head. Maddie's mother Kelly, who'd been concentrating on the slippery, unsealed road ahead, paused and pulled over.

Mother and daughter, who owned horses themselves, looked down at the riverflat closely. The foal seemed to have broken its leg, Kelly thought. She turned the four-wheel-drive and headed down into the paddock. Maddie jumped out of the car as soon as they reached the foal and covered it with her raincoat. Even under the coat it looked pitiably thin. To Kelly it looked like an animal that was dying. 'We've got to find someone with a gun who can put this poor little creature out of his misery,' she said.

By now it was hailing. Maddie, shivering visibly, said she would stay with the foal while her mother drove down the road to find someone who could help. Kelly knew it wouldn't be hard to get hold of someone with a gun in an area full of recreational shooters and farmers. It took her perhaps ten minutes to negotiate the dirt road down to the bitumen road that goes to Omeo; long enough for her husband Tom, driving ahead of them in another car with daughter Katie, to notice that his wife was no longer behind him. Tom waited, then turned back, worried that Kelly had gone off the side of the winding mountain road. He called another neighbour, Cath, in the valley to see if she could see his wife's four-wheel-drive before turning to drive back. Kelly meanwhile had found a man with a gun.

The three cars converged on the riverflat and the foal: the neighbour Cath; Tom, relieved that his wife and daughter were safe; and Kelly and the shooter. Maddie dreaded what was to come but she knew her mother had the horse's welfare at heart. But when Kelly took the raincoat off the foal she could see it wasn't suffering from a broken leg—it was malnourished.

'What the hell are we going to do?' she asked.

The Crombies then realised that they'd seen the foal before—with Helen Packer only weeks previously. Cath recognised it too as 'Clancy', orphaned after his mother died during a severe episode of colic. 'Packer', as Helen is known, had noticed the mare's sides had become distended when she'd been eating some lucerne hay one day. The mare was still feeding but there was something not quite right, she thought. The horse's gut was blowing up like a balloon; she had gas colic. Packer called the vet. As she was waiting for him to arrive she checked Molly's gums and saw that they had turned white. Even as she held the mare's muzzle in her hands, Molly slumped down dead. Clancy tried to suckle her, nudging his dead mother for a drink then lay down next to her.

Clancy was nearly two months old when he was orphaned and big enough to be on his own. Packer had charted his progress in her 'horse book'; he seemed to be doing well and was gaining condition. He was a funny little foal, who used to stand with his tongue hanging out, looking comical. She was satisfied that he was strong enough to cope when she left her horses with someone overseeing the property to go interstate two weeks later, setting up a special feeding system for him.

But a combination of factors including a bout of extreme weather caused Clancy to lose condition quickly and for his immune system to crash. In a short space of time his health deteriorated rapidly, his plight unnoticed as he moved among the large herd of mares and foals. The Crombies were looking at a shadow of the foal they'd met several weeks earlier.

Kelly looked down at him again. The foal was obviously dying; why prolong its suffering?

She noticed Maddie looking at her.

'Mummy, you can't shoot him,' she said. 'He's not ready to die.'

They tried to lift the foal to his feet but he crumpled down.

'Do you want to take him home with us?' Tom suggested.

And with that, the family swung into action.

They lifted Clancy up and put him into Cath's ute, covering him with some Drizabones while they thought of the next move.

'Where the heck are we going to find a horse float?' asked Kelly.

The owners of the Blue Duck Inn nearby were happy to let them borrow their float. They lined its floor with some old mattresses. Cath provided a heat-retaining 'space blanket', used for hypothermia in people. They shifted the foal onto the mattress. He was prostrate by now, unable even to lift his head. Kelly called her sister-in-law, Amanda, who lived on the farm with them. 'Get a vet to meet us when we arrive,' she said.

The trip to Metung would take about three hours, up, around and down the mountains from Angler's Rest through Omeo then down towards the coast. The road wound like a serpent through those mountains, dropping down sharply to the riverflat below and there were no shortcuts. Kelly was terrified the foal would die on the way, that the shock and movement as the float swayed on the curves and bends would be too much for his frail body. She wanted to sit in the float with him but Tom and the others at the inn talked her out of it, insisting that it was too dangerous.

They travelled back in convoy, Kelly getting Tom to pull over on the way so she could check that the foal was alive. She says she fully expected to open the tailgate at the other end and find a dead horse.

The foal made it to Metung but the vet wasn't there when they arrived home. Kelly was frantic—they'd driven all that way only to find that the veterinary care they needed so urgently wasn't available. When the vet arrived half an hour later after finishing another call, he told them that Clancy was so dehydrated that he couldn't get an IV line into his limp throat to administer the electrolytes he needed. Electrolytes keep the body hydrated so muscles and nerves can function. The vet left the family with an electrolyte solution, although Kelly could tell he was thinking that the foal was going to die. Clancy was hypothermic and a blood sample would later show that he was severely anaemic and metabolically critically ill. Kelly and Tom knew they had to get fluids into him urgently. They sat on the floor of the float with Clancy wedged between them all night, taking turns to dribble the electrolyte mixture into the side of his mouth. Clancy only managed to drink about a cupful, but it was a start.

Tom turned one of their open horse shelters into a stable, erecting wooden boarding to keep the weather out, and shovelling in sand to give it a floor. They put a mattress on top of the sand then moved the foal, still lying on his side, into the stable. They turned him over at intervals to stop him getting sores, dribbling more electrolyte solution into his mouth at every opportunity.

But Kelly discovered that it wasn't just a matter of sustenance and that no amount of fluids or food alone would save him—Clancy was riddled with worms. Red strongyle worms had taken hold of his weakened body. Without his mother's milk Clancy's immune system had been compromised and the worms were latching onto the inside of his gut, breeding in the gut wall and robbing him of nutrition. He was passing manure that looked like red spaghetti. Kelly and the others in the family scraped it up and disposed of it immediately to avoid infecting their seven horses.

Kelly sought help from an equine specialist the next day, who talked her through it on the phone.

'I'm going to be straight with you,' the specialist said. 'This horse needs intensive care. He's probably not going to survive but this is what you should try to do . . .'

She recommended drenching the foal, and prescribed a list of medicines, herbal preparations, feed and supplements 'as long as your arm'. The Crombies drove around collecting the items on the list. A local feed supplier donated some pellets and lucerne. A special diet was formulated. But when she had it all in front of her, Kelly wondered how they would be able to get it into him. She put the various supplements in a coffee grinder and made a mash of it, mixing it in with some lucerne chaff. They fed it to Clancy by the spoonful, as much as he could swallow at any one time.

The family sat with him on shifts over the next six or seven days—Kelly from ten in the morning to six; then Amanda; then Amanda's partner Anton, a chef who worked at night, doing the midnight shift when he got home. Kelly would join him in the stable at night sometimes, unable to sleep.

After a few days she tried to stand the foal on his legs. He buckled at the knees and fell down. She kept trying every day. Then, on the sixth day as she went to lift him, Clancy stood up, nearly knocking her over. Kelly didn't know whether to laugh or cry. But she knew that he had pulled through.

The little chestnut recovered well. When he was strong enough to walk around they matched him with an old pony of theirs, Zeke, a gelding that looked after him as zealously as any mare. At first the pair was put in a small paddock near the house together, Zeke never more than a few steps away from the foal. Several weeks later Kelly put them in with the other horses, Zeke flanking Clancy and warding off any horse that got too close.

Kelly stopped counting the cost of the vet bills, medicines and feed but knows it ran into the thousands of dollars. She didn't care. When her friend and neighbour Packer returned from her trip, she was shocked to hear what had happened to Clancy, amazed that his health had deteriorated so sharply. But the first thing she said was, 'I want Maddie to have him.'

Kelly agreed—the family had invested too much love in the foal to be able to give him back. The kids had become too involved and even Tom, a 'big burly bloke', had been affected by Clancy. In the early days of Clancy's recovery Tom would fall asleep with his arms wrapped round the foal.

Clancy's now four, healthy, robust and beautiful. He's a stocky, well-proportioned horse. His father, Captain, is a handsome Friesian–warmblood cross and mother Molly was part Clydesdale. He's smaller than his older full brother, though, perhaps because of his hard start in life.

Maddie became responsible for his training, and Clancy, who'd been so intensively handled, gave her no trouble. Maddie saddled and backed him without any hitch, and he's responded to every challenge that they've given him since. Clancy takes everything in his stride—carrying a pack, walking through bog moss and streams, jumping over logs, ignoring cracking whips and noisy motorbikes. They haven't found anything that rattles him yet. He stands waiting for Maddie at the fence in the morning as the other horses graze. He throws his head into a halter. He loves to be around people and loves being ridden. He's a joy who's repaid their effort to save him in spades, says Kelly.

They've got him for life.

PART X

There's always one

32

BETH AND GYPSY

Beth Mackay (nee Densley) was born in 1918 in the era when horses were used on farms for ploughing and rounding up cattle and sheep, for riding to school, or going to church, the store or the beach pulling a jinker or buggy. But she still rode way beyond when that era ended. After a childhood and young adulthood of using horses for work, Beth discovered recreational riding. In fact, if it weren't for the death of her aged mare when Beth was 74, she would have been riding longer still—after all, she only gave up playing competitive tennis at 86.

Her family lived on a farm in the rolling green Strzelecki foothills of South Gippsland, or as she calls it 'up and down country'. The use of horses had long been vital to life in the area. In her grandmother's case it could mean the difference between life and death: Mary Densley was the local midwife and travelled to homes in the district on horseback to deliver babies. Because there were no telephones, Mary could answer a knock on the door at any time to find the husband of a woman in labour who had come to get her and who would ride back with her to their farm. Beth never knew her grandmother because she died two weeks before she was born but she's heard about her work. There are people in the area who have relatives with birth certificates recording the presence of a midwife called Mary Densley.

Beth grew up on a mixed farm of several hundred acres where her father raised sheep, then later dairy cows. The farm was divided into two large paddocks, the 'day paddock' and 'night paddock'. Members of the family riding horses brought the cows to the dairy from these paddocks twice a day. She rode to school in Dalyston, 4 miles or so away, double-dinking with her older brother Frank on their pony, Dot. Dot was presumed to have been a mine pony at the coalmine at Wonthaggi, the town further on from Dalyston, as she was already mature when she came to the Densleys and was aged by the time Beth left school at fourteen. When they arrived at school they'd turn her loose in what was called the 'Pony Paddock' along with the other children's mounts and the sewing mistress's horse, if she were there that day. Dot was a good little pony but it was always a thrill for Beth to be able to ride her older brother Arthur's horse to the blacksmith's to be shod—a big horse was an exciting change.

The family had a two-horse buggy, a double-seater with four wheels, to drive into Wonthaggi to shop, visit friends or go to the doctor—although you had to be *really* sick for that, Beth says. It was pulled by ponies that were 'good little movers', though Beth always preferred to ride if she had the chance. There was a stable in town where you could leave your buggy and horses while you did the shopping, and troughs in the main street to water them, long since removed.

Beth recalls cars appearing in the district when she was about seven in 1925: T Fords. Not everyone could afford a car so continued to use horses, and even when they did buy motorised vehicles horses were needed as a back-up. Beth remembers her parents' car getting bogged on the highway, and having to go home and harness the horses to pull it out. The highway was then a dirt track full of ruts caused by the horsedrawn vehicles.

Horses had to be versatile then; everyone in the family rode everyone else's horse and they rarely played up because they were always in work. But Beth's brother Frank had a horse that proved an exception. If the mare was in constant work she was fine, but if two or three days passed when she wasn't in harness she'd either refuse to move or move sideways. Frank would use a whip on her and she'd still do it. One day he did his 'na-na' properly and let loose a torrent of swearing, for which he was thoroughly reprimanded by his mother. Little Dot would buck but only if you put your heels back on her flanks, Beth says, something the boys did for fun.

Beth fell off often. There were plenty of rabbit holes and it was only a matter of time before your pony put its foot in one of them and you took a tumble. No broken bones, though. Her father owned another farm in Kilcunda, a few miles down the road, and Beth and her sister Beryl used to ride down to check on the cattle and bring back any cows that were ready to calve. It was a job she didn't enjoy one bit. It was difficult to separate the cow you wanted from the herd and get it through the gate without the other cattle getting on the road. The cow invariably resisted leaving the herd and, even if you got it out, would try to get back through the fence again. Sometimes a cow would already have calved and you'd have to find the calf.

Beth married Alex, a farmer she'd met at the Woolamai dances, in 1939. The couple moved out of the area when Alex got work on a farm. Beth missed her family's horses when they moved: Molly, the black hack; Venus, the grey Arab; and Bessy, the bay pony the family found in one of their paddocks one day. Bessy's arrival was always a mystery. There'd been a big strike at the mines at Wonthaggi and the family could only imagine that someone couldn't afford to feed her when they weren't working and had put her there for a while. After some time they decided to catch her and found out that she was an excellent jinker horse. No one ever came to claim her and Bessy stayed.

Beth and Alex had horses of their own when they later lived on his family's farm at Woolamai, using them mainly to round up cattle. It was by chance that Beth came to own Gypsy, her best horse—and it was long after she needed one. The couple had been looking for another mount and after an unsuccessful attempt at buying one from the Dandenong Market—it bucked when Alex brought it home then ran him into a barbed wire fence—they decided to breed from a mare they'd been given, Girlie. The result was a bay foal with a white blaze called Gypsy, which Beth had broken in when Gypsy was old enough.

And that began a new era of riding for Beth, then 52—she was riding solely for the joy of it. She formed an adult riding club with a friend for local riders who'd meet at each other's houses, bring their own lunch, have a cuppa and go for a ride. They trucked their horses to the punt that goes to French Island one Easter and stayed there, exploring the island. They rode to the ocean beach at Cape Woolamai and galloped along the water's edge. They walked their horses at low tide over to Snake Island further up the coast, though Gypsy, being 14 hands or so and smaller than the other horses, found herself swimming for awhile. They camped out in the horse float on a ride in the Victorian high country. They raced each other at the Woolamai racecourse after the official country races were over, just for fun.

Gypsy was a great little walker, keeping ahead of much taller horses. Beth noticed a man in the riding club trying to keep up with her one day, urging his horse to walk out faster. He was puzzled. 'I thought my horse was pretty good, how did you do that?' he asked. Beth replied that she didn't, that the horse did it by herself. She says Gypsy's energetic walk was helped by the times when the two of them were heading for home after a ride and Gypsy would speed up at a walk ready to take off into a dash for home the moment Beth loosened the reins. Beth would sometimes take the lead on the club's rides when they needed someone to go through a creek first or, as happened once, a series of big puddles in a laneway that the other horses were baulking at. 'Make a clearing, I'll take Gypsy through,' said Beth. 'They didn't want to get their feet wet!' she jokes. She'd power up hills as other horses paused to stop for a breather.

Gypsy rarely caused Beth worry of any kind and she never fell off her. She came close once when the mare put her foot in a hole and came down on her knees and nose, causing it to bleed. Beth says the stock saddle she rode in stopped her falling off. Gypsy became lame once when they rode into a boggy patch, sinking into the soft silty soils so that the mare couldn't get out and Beth couldn't get off. Gave her a real fright, she says. Gypsy got herself out but injured a fetlock on the way. Beth took her to a chap who was good with horses; he advised her to confine and rest the mare for three weeks. Alex, wondering why it was taking so long for Gypsy to recover when she had no visible injuries, tested her fetlock by giving it a squeeze—and was bitten for his troubles.

Gypsy was well-behaved but not everyone could handle her. Beth only ever let one of her granddaughters ride her. 'If you ever want a home for Gypsy, Gran . . .' the granddaughter hinted broadly one day.

'By the time I'm finished with Gypsy no one else will be able to have her!' replied Beth, a bright, firm-minded woman.

Gypsy did live to an old age, though not as old at Alex's mount Ginger, who died at 40.

One day in July 1992, when Gypsy was approaching 25, Beth noticed that her stomach was distended. She was sure she couldn't be pregnant but decided to check anyhow. Beth asked the friend who knew about horses to have a look at her. He confirmed that she wasn't pregnant but couldn't shed light on the distended stomach.

She called in the local vet, who examined Gypsy and said she had a growth that would only get bigger. Beth had ridden Gypsy not long before so it came as a shock. She didn't want Gypsy to suffer so she decided to have her put down. It was 11 July 1992. The date's written in the back of an old diary.

'She was the best horse I ever had to ride.' More than two decades later the thought of it still causes Beth's pale blue eyes to go misty.

Beth went on a trail ride some time after that with two of her daughters and friends, hiring horses from a man at Powlett River. When the man asked what sort of horses the women wanted, they all said things like, 'Give me the quietest horse,' but Beth asked for a good walker. But that's all they did the whole ride—walk. 'Thirty dollars!' she exclaims. 'What a waste of money!' She never rode again.

33

A HORSE CALLED SPOOK

Helen Packer has ridden more horses than she can remember in the time since she was a jillaroo in the late 1970s. Owns 70 horses now in her trail riding and trekking outfit, give or take a few. She's ridden handy stock horses, horses that are real characters and some that are memorable for all the wrong reasons. She says that rarely—maybe once in a lifetime—you come across one that connects with you. A horse that is in tune with you in a way you can't quite put it into words. Spook was such a horse.

'Packer', as she's called by those who know her and her tough, no-nonsense ways, first worked with horses as a jillaroo on a sheep station in the Pilbara. She was on vacation, studying science at university in Hobart at the time, but decided she liked the Western Australian outback better. Loved the wide, open country with its red dirt, endless horizons and big skies. Her soul still sighs whenever she sees that red dirt. She worked on Yanrey, a property the size of a small European country, 750 000 acres. To the west from the homestead was dune and spinifex country, one paddock so vast that it was called 'Siberia'. To the northeast was open river land and the Ashburton River, then just a series of pools lined by tall, shady trees. Packer arrived after five years of drought. The way of life was harsh, nature almost cruel, and the people and animals had to be tough to survive.

She'll never forget Baldy, the horse she was given to work the livestock with.

'Watch out, he kicks,' her boss, station manager Tom Alston, said as he handed over the reins of the bay gelding with the broad white blaze. It was something of an understatement. Baldy's ferocious double-barrelled kick could send man and beast flying. He was known for it. Once the boss was booted clear out of a float when he was loading Baldy to drive out to a muster early one morning—and Tom must have weighed 18 stone. It wasn't really the horse's fault, explains Packer. He'd been trained to kick out at scrub cattle as a means of self-defence. Scrub cattle are wild, they'll go you as soon as look at you. You never approach scrub cattle on foot. Packer developed a pulley system for loading Baldy on the float, tugging the horse up the ramp and inside with a rope hooked around the front bar so she could stand at a safe distance from his legs.

From Yanrey she moved south to the sheep country of Gascoyne, just below the Pilbara. Flat as far as the eye could see, the landscape dotted by spinifex and saltbush. Much of the work here was done on motorbikes, checking windmills from dawn to dusk. She then had a stint as a 'ski bum' in America, Austria and France in the mid-80s, which indirectly led her back to riding and to a life working full-time with horses.

She'd shared accommodation in Austria with a fellow skier, a woman called Billie Kelly who was from one of the well-known families of the Victorian high country. Billie invited Packer out mustering cattle with her father and uncle when they were both back in Australia in 1989. It was late April, the end of the summer grazing period, and the weather was foul: foggy, damp, wet and made you cold to the bones. But Packer loved it. She was back in the saddle.

She worked for a man taking out groups of people on trail rides; a 'colourful character' with whom she didn't see eye to eye, Packer says. She decided to do it for herself, got together a small group of horses with the help of another ride operator and rented land at Dinner Plain, in the mountains, gathering more horses as business grew. Spook was to become her lead horse.

Spook was a reject from Cobungra Station. A big grey horse, he had a dash of something 'high-stepping' in him, perhaps Andalusian, the Spanish horse. He 'jig-jogged', as Packer put it. He'd been bought by Cobungra to work stock but wasn't suited to it; the jogging wasn't conducive to calming cattle. But his name came from the way he looked—pale as a ghost gum—not because he spooked. Packer handed over 350 dollars and rode out of the gates of Cobungra and 20 kilometres or so back to Dinner Plain on what turned out to be the best horse she's ever had.

The pair clicked from the start. There was nothing pretty about the 'boof-headed' horse, except perhaps his full white tail, but he was brave, solid and strong. Despite his fancy footwork, he'd go anywhere. In time he stopped jogging with Packer. He became the horse that would nicker to her from a group of horses in the paddock. Packer could almost whistle him up out of the herd—Spook *wanted* to go on rides. He had all the best qualities of a lead horse: independence, boldness and total trust in the rider. You can always spot a lead horse, she says. They're the ones standing by themselves without caring if there are other horses around them or not. You can take a lead horse up and down the line on a trail ride and they won't care if they're first, in the middle or at the back. They trust you, work with you and you never have to do too much to get them through a tricky situation. You can trust your lead horse to carry someone else—if a rider is nervous about the horse they're on, you can always say, 'Here, take this one.'

Packer used to pick up work elsewhere while she was building up her trail riding business—her mother had long been pressing her to get a 'real job'—and for two years she ran the Brandy Creek restaurant and bar at Dinner Plain. One spring night after the meals were finished, she was enjoying a chat with a group of diners at the bar. Talk was flowing freely when the front doors parted and Spook walked in with two young women aboard. The women, who worked for Packer, looked around and giggled, cans of Bundaberg rum in hand.

'What are you doing on Spook?' Packer asked.

'Well, he was stupid enough to come up to the gate to be caught so we decided to ride him in,' one of the women said, burping rum. Spook stood on the slate floor at the bar looking boof-headed and unfazed.

The chef, however, was flummoxed. He came out of the kitchen, saw the horse in the bar at the front and scowled.

'You've got a horse in the bar,' he said to Packer.

'I can see that,' she said.

'You're not allowed to have a horse in the bar,' he said.

'Who says? He's not in the kitchen. The food's done for the night so it has nothing to do with the food.'

'I won't work in a place where there's a horse,' declared the chef.

'Well, I can cook,' said Packer, thinking it was October, the ski season had ended and there wasn't as much call for meals.

The chef looked at her with narrowed eyes, huffed and walked out.

As it happened, the restaurant was mostly filled by a group of farriers and jockeys who got together each year to come up to the high country to ride with Packer. They thought it was amusing that someone would be so uppity about having a horse inside. The chef came back three weeks later and Packer gave him his job back. 'But I can't guarantee that there won't be another horse in here again,' she said.

Packer bought Willows Retreat at Anglers Rest in 1999, her remote horseriding and accommodation complex. She continued to lease the land at Dinner Plain and ran rides from both places for several years. She and Spook travelled many miles together through the high country, bush-bashing as Packer sought out trails, and leading people on short rides and five-day treks. It wasn't until she put another rider on him, though, that she realised just how well they had gelled.

'Oggie', a British friend of hers, was a seasoned rider. He was a big-boned man and she decided to give him Spook, a big-boned horse. Oggie was frightfully English, hunted, and spoke with a stutter. They set off from the yards towards the high plains. Spook started jig-jogging. Packer laughed as she watched the big man bumping around. 'He doesn't do that with me!' she said.

Oggie got more and more frustrated as Spook jogged about, refusing to settle. After they had ridden for awhile he exploded, 'This f-f-f-f-f-fucking horse! Give me another f-f-f-f-f-fucking horse!'

Packer eventually moved her horses to Anglers Rest permanently in 2005. They roam across 350 acres of bush, meadow and granite boulder country that weaves in and out of several hills, with kilometres of riding trails beyond. She'd been there a few years when, one hot January day, some anglers staying in one of her timber cottages came up to tell her that they'd just seen a big grey horse drop down, dead. Packer knew that Spook had cancer—he'd had black cysts under his tail for as long as she'd owned him but the vet said there was no point treating them as there could be any number more inside. Spook had been standing this day with some other horses in the shade of a shearing shed when he collapsed.

'There's a horse down there that's just dropped dead,' the anglers said, somewhat incredulously.

'Fantastic,' said Packer. 'That's old Spook.'

The men looked at her, mortified. Packer explained about the cancer. It was a quick, kind way of going, she said. Rather than suffering, he'd been standing in the shade swishing flies and just fell over. He'd lived his life and done her well.

'But I missed him. He was a bloody handy horse.'

Spook was laid to rest between the granite boulders down the slope from Mount Ned in what's now known as Spook's Gully.

There have been other lead horses since Spook. Right now, Packer's grooming young Sox, who finds his way into the kitchen, scrambles through the bush with ease and is a bit of a clown. He'll be able to give old TJ, her experienced lead horse, a break. But there are still times when she'll be riding past the boulders beneath Mount Ned and think of the one who will always be her best lead horse.

34

ZELIE BULLEN AND BULLET

Long before Zelie Bullen became famous internationally as an animal trainer on the movie *War Horse*, she had an encounter with a horse that was to unfold into what she calls her own real-life love story.

The story begins on a wintry day in 1997 when Zelie Thompson, as she was then, was living and working on the Gold Coast hinterland as a polo groom. She was driving home one afternoon after a polo match at Beaudesert with her boss, Wayne, a professional player, when he told her they were going to stop off to see a horse dealer about a mare. It was a horribly wet day and Zelie was happy to sit in the horse truck waiting while Wayne talked to the dealer and rode the mare. As she sat, she looked across the post-and-rail yards past the dealer and Wayne. A chestnut horse was standing in another yard, knee deep in mud, looking at her in a way that she found compelling.

She got out of the truck. The horse pricked up his ears and walked over to the fence. Zelie bowed her head in the rain and made her way to the yard, patting the horse's damp neck through the fence, then walked to where Wayne and the dealer were finalising the deal for the bay mare.

'What's that horse?' she asked.

'Argh, no, you don't want that one, luv, he's no good, he's off to the doggers,' the dealer replied.

'What's wrong with him?'

'I dunno. He fell off a truck, off the tailgate. He's a racehorse from Brazil.'

'How much do you want for him?'

'No, he's no good,' the man said.

But Zelie persisted and persuaded the dealer to sell her the horse. She led the gelding over to the truck and loaded him on with the mare. Wayne shook his head jovially and muttered something about 'a waste of money' as they drove off.

The horse hadn't been cheap—at 750 dollars he cost more than any horse she'd bought before and was the same price as the mare. Zelie knows it sounds 'airy-fairy' and couldn't tell you what it was, but there was just something about that horse. Besides which, she told herself, he'd been a racehorse good enough to import from Brazil so he must have had promise once.

She put the gelding in a yard at the polo property, leaving him alone with a biscuit of hay. It wasn't until the next day when she went back to feed him and clean him up that Zelie noticed that he had four white stockings and that his blaze had a chunk the shape of a bullet missing in it. She called him Bullet.

Bullet was a beauty, a barrel-chested thoroughbred that looked more like a quarter horse, aged two. He was broken in but apparently had never raced in Australia. After letting him settle in for few days, Zelie decided to test out her new horse, got on, and walked him round a yard, noticing happily that he had a soft mouth. But when she asked Wayne to ride him and Wayne took him out of the yard, Bullet mucked about, fidgeting and moving sideways. Wayne, thinking he was going to buck, took him for a quick canter up a hill. Superficially the horse seemed sound.

But Zelie noticed after she took him home that Bullet's chalky-white feet had abscessed from his time standing in the mud. He was by now slightly lame—but not lame enough to stop him taking off as soon as she released him into her 7-acre paddock. Zelie had only just bought the property and house, the first she'd owned, and hadn't built any small yards. Bullet, perhaps always having been stabled, panicked at the sight of the open space and ran headlong straight into a fence, cutting his face.

As Zelie decided whether or not to call the vet over the small cut, she suddenly wondered about all the other things that might be wrong with Bullet, what she should be doing for him and about the vet's bills this might involve. All she'd seen at the dealer's yard that day was an animal in need and one that was looking at her with what she felt was an intensity she couldn't ignore. She decided not to get the vet out but instead called an experienced vet nurse who was a friend, to come and help her stitch the cut. Then she gave Bullet time off to spell.

Bullet had one injury after another in the next two years, mostly hoof problems. Zelie kept trying him out to see if he was sound enough to ride, then telling herself when he wasn't that she had enough grass to keep him and she quite liked him, trying not to think what would happen if he turned out to be unrideable. Fortunately the paddock around the house was lush with grass so feed wasn't a problem. She'd struggled to feed her horses as a child growing up in a single-parent family in Darlington, a small town near Perth, walking its streets asking people if she could fence off a patch around their house for her pony to graze on in return for clearing their paddocks of manure or other odd jobs. Zelie's father had left them when she was a baby and the family couldn't manage the expense of hand-feeding several horses, yet when Zelie outgrew her first pony her mother didn't make her sell her; she knew what the pony meant to Zelie, a sensitive soul.

She had other matters to occupy herself at the time that she was spelling Bullet. Her days were filled working and also training as a stunt performer, going to as many courses as she could: kickboxing, gym, taekwondo, stunt driving, scuba diving or learning high falls. Wayne knew and accepted her priorities; firstly she was a professional stunt performer and secondly she groomed polo horses. He let her have time off between caring for and exercising the polo horses to train and to take roles on various film and TV productions, several of them at Warner Brothers Studios on the Gold Coast.

When Zelie next rode Bullet in 1999 he was still acting like a 'naive racehorse' despite the time he'd had to acclimatise to life in the country. She was trying to get him across a creek one day when Bullet panicked and leapt over the whole watercourse, landing in the bush opposite and spearing his hoof with a stick. The stick pierced his coronary band and went into the pedal bone of his foot. The injury required surgery and meant more spelling. Zelie couldn't believe it—she'd only ridden him perhaps twenty times since she'd bought him. 'I don't know what it is about you, mate,' she said to the horse when he came home from the vet, 'but I'm going to hang in there with you.'

The bone took a year to repair. When it did Zelie starting working Bullet in earnest, happy to be riding him at last. It had been three years since she'd bought him; Bullet was now aged five. But when she was doing some circle work on him one day, she noticed that he seemed stiff. 'Wow, he's got such bad flexion,' she thought, 'I'm going to get him a massage.'

The horse masseur came and worked her hands over him, then stopped what she was doing, turned to Zelie and said, 'I'm not sure if I can help this horse, Zelie. There's something wrong with his neck. Can I just call a friend in?'

The masseur called a horse chiropractor, who examined Bullet and delivered his diagnosis: three of Bullet's vertebrae were fused together. Apparently, he'd broken his neck falling off the horse truck.

Zelie was deflated. Her older sister Freda came to visit at the time and Zelie told her about the fused vertebrae. 'What am I going to do now?' she pleaded.

Freda looked at the horse standing with them, his soft eyes, and the way he looked at Zelie. 'You can't give up on him,' she said.

Zelie persevered. Bullet wouldn't ever be able to canter in small circles but he could still be ridden without being in pain. She started to educate him. As she spent more time with him, grooming, training and riding him, the relationship shifted. Bullet had been more interested in her small herd of horses when Zelie first brought him to her house but he started to warm to her. She noticed that even when he was with the herd that Bullet pricked up his ears whenever he saw her and that if she was doing something else around him he would always be looking to see where she was.

At the time, she was learning trick riding from a top instructor in America, Tad Griffith, and would return from each trip with exercises he'd given her to practise with her horses, mainly with Bullet. The showy chestnut took to it. He was gentle and forgiving as she practised, never kicking her as Zelie moved over, under and around him, fell off or fell on any part of him by accident. Bullet trusted her; she put her faith in him.

She perfected the stunts, standing upright on him as Bullet tore along the sandy arena, jumping on and off him, hitting the ground and bouncing back up into the saddle again, swinging by one leg upside down from the saddle in the Cossack Drag, and in the most elaborate move, passing down his shoulder, then under his neck and then up the other side and back into the saddle. Bullet galloped hard and fast in a straight line, always concentrating on what he was doing.

Trick riders practise their moves 'a thousand times' at a standstill, then at a walk, then get faster—but they work best at a gallop. There's a moment in the cadence of the gait when all the horse's feet are off the ground and it's airborne. The trick rider aims to hit the ground just before this moment—the 'sweet spot'—and fly back up again, Zelie explains. Miss it and things can get ugly, messy or dangerous. You have to be able to trust your horse's every stride. And she did.

Horse and rider started to perform the stunts they'd been rehearsing at small rodeos and horse shows, then at bigger agricultural shows. Zelie, long dark hair streaming behind her, dazzled the audiences with her athletic displays. The arenas and the audiences got bigger. Zelie was invited to perform as a precision rider in the opening ceremony of the 2000 Sydney Olympics but had to choose a horse other than Bullet—horses with white markings weren't allowed because they'd stand out too much. Thousands of people Australia-wide watched Bullet and Zelie gallop around the arena as part of 'The Man from Snowy River' spectacular in 2001.

Her proudest moment came in an act she and Bullet performed with her friend Deb Brennan and with their instructor and mentor Tad Griffith at Equitana International's trick riding event in Brisbane 2002. Riding in a live show alongside Tad for the first time was a dream come true for Zelie, leaving her with a feeling of tremendous accomplishment and pride in both herself and in Bullet.

Zelie was by now realising an even longer-held dream: to become an animal trainer. It was like a 'calling', she says. She'd had an affinity with animals for as long as she could remember, particularly with horses and dogs, and, it's said, has a 'gift' for communicating with them. She often sought solace in the animals around her during the dark times in her life. As a young adult, within a space of five years, Zelie lost a sister, the father she'd just been reunited with, the man she was going to marry, and a close friend to suicide. There were times in her life when she would bury her head in the warm neck of a horse and weep.

She grew to understand how and why horses behaved as she spent time with her 'herd', recognising what their body language and facial expressions meant. As a girl she'd taught her and her girlfriends' horses tricks, getting them to flick an ear, flare their nostrils, paw the ground or rear. 'Very bad habits to teach ponies!' she says now. As a young adult she spent as much time as she could around the best trainers she could find, eventually working for some of them. Finally she became an animal trainer in her own right. Between 1996 and 1999 Zelie worked on more than 30 productions in Australia and overseas with all kinds of animals, as well as her stunt work and live performances. She was flourishing.

In 2000 she met Craig Bullen (of the circus dynasty) and teamed up with him to work on movies including *Mask of Zorro* (2004), *Racing Stripes* (2005) and *Charlotte's Web* (2005). The couple married in 2005 and had a baby, a boy called Colt, in 2006. They were employed on Steven Spielberg's *War Horse*, the wartime movie about a boy's love for a horse, in 2010. Zelie formed a strong bond with the horse she worked most closely with on that film, Abraham, a Dutch warmblood who acted in the more sensitive scenes, at liberty. (There were nine horses playing the central horse 'Joey' in the film.) She was promised a horse from the set and wanted to keep Abraham but the owner reneged on the deal and the horse stayed in England. She's still hopeful about being reunited with Abraham one day. But she calls Bullet 'my true love story'.

There's more to Bullet than being Zelie's best performer. Bullet turned out to be a beautifully natured horse: gentle, friendly, obliging and loyal. 'I don't know if he was like that all along or if he went, "You know what, you're going to look after me, I'm going to look after you",' she says. 'Maybe I got lucky. I love all my horses—we've got eleven horses and eight ponies—but I adore the ground he walks on.'

Bullet became the horse Zelie trusted enough to put anyone on: friends, beginner riders, people with disabilities or elderly actors who needed to feel confident in the saddle before they rode in a film. Her son Colt learned to ride on Bullet, starting as a baby, holding on to the front of the saddle. Zelie says she wouldn't have put her son on *any* horse. Bullet, at 15.2 hands, is the wrong height for a beginner child yet was perfect for Colt, who rode him independently at four and was cantering him when he was five years old. Visitors to the Bullens' property in the picturesque Canungra Valley are invariably given Bullet to ride, no matter how experienced a rider they are. 'I say, "Don't lean on his mouth" and let them go and Bullet really doesn't care.' He's become a favourite with her family and girlfriends, who sigh fondly whenever she mentions his name.

A squabble broke out when all six of her bridesmaids wanted to ride Bullet at her wedding. Zelie was riding another of her special horses, an Andalusian mare named J'dore, because she felt a white horse would show off her red raw-silk wedding dress the most. Freda won, insisting that she was the most nervous rider and that if she couldn't ride Bullet she'd be coming along on foot. One of the groom's party was asked to saddle Bullet on the big day while the bride got ready and inadvertently did the girth on the Western saddle up too tight. 'I always need a tight girth as a trick rider but I've never seen one as tight as that,' says Zelie. The girths on Western saddles tighten more easily than on other saddles. Bullet started jogging uncomfortably as the bridal party headed towards the hitching rail surrounded by hay bales where Zelie and Craig were due to get married.

'What have you been feeding Bullet?' asked Freda, surprised at the horse's behaviour.

Then he started doing piaffe and passages. '*I'm* the bride—why is he making Freda look good!' Zelie joked, watching the fancy footwork. Then she became disappointed at her favourite horse's apparent misbehaviour. 'Why now, Bullet? Any other time you're perfect!'

As they were about to set up for the photo session, however, she realised there must have been something wrong with him and asked everyone to stop what they were doing. Stiletto-heeled boots sinking in the ground as she checked the saddle and loosened the girth. Bullet stopped jogging.

He can be a 'rat' at other times, though, when he wants to be with the other horses, and can behave like a stallion when he's around the mares on the property. Over the years, Bullet has moved up the pecking order and is boss in the paddock and 'sooks' if he's taken out of it unexpectedly and away from the other horses. 'He has a mile of respect and love for me,' says the woman so keenly attuned to horse behaviour, 'but he's got an important horse existence, too.'

Bullet has a good life and appears contented. But Zelie can see him getting older now. His bottom lip has started to droop a little when he's resting, his mane is thinning and a few grey flecks have appeared on his nose and under his eyes. Zelie worried that a bald patch that appeared on his rump might be something sinister but the vet assured her it was just old age. 'He could go for another ten years, couldn't he!' she says with hope. Indeed, he could. The story of Bullet, the horse that could so easily have lost his life, goes on.

35

THE PACT

When Graham Payne bought his first horse at 45, he knew very little about horses. Can barely recall riding before then. He knew he wanted to buy a riding horse, preferably one that was also trained to harness, but beyond that he was going to let the horse pick him. Reckons that's the best way to start a relationship with one.

Grae, as he's known, was living in Warrandyte, a semi-rural suburb on Melbourne's outskirts with plenty of places to ride, and he owned an acre of land in Shannonvale in Victoria's high country. He started casting around. The horse in the first ad was ugly—a stocky thing that didn't seem to like him. He didn't like the look of the second horse, a Morgan, either. But when he went out to see the third horse—Samson, a 16.2 hand Percheron–quarter horse cross—he was stunned.

Samson was a strikingly handsome bay with a commanding presence. He was broken to harness, too, which was a big consideration—Grae had a dream about buying an old milk lorry and taking hayrides along the riverbank near where he lived. The woman selling Samson took Grae and his daughter for a spin in her 'two-wheel brake', a small cart a bit heavier than a sulky. Grae then rode Samson, who did exactly what he asked of him, filling him with confidence. It was getting better. The two adults were standing together talking about whether or not Grae would buy Samson when the big horse gave him a gentle nudge in the back; the deal was clinched.

But first the woman, who had owned Samson for some time, wanted Grae to make a pact with her. 'I want you to promise that if he ever gets in any pain that you'll put him down and won't let him suffer,' she said. Grae promised he would do this and took his horse home.

He vowed to himself that this was going to be a relationship based on trust and respect. Samson was eleven years old, clearly had a mind of his own and could assert himself if he chose to—he was not a horse to be pushed around. He'd been trained for polocrosse and had become very successful at shoving the other horses out of the way during games. But right from the start, Grae had a sixth sense that even with so little experience he'd be able to handle the big, bossy horse. He gave Samson the respect he deserved, and Samson taught him about horses.

Not long after, Grae found the milk lorry he'd dreamed of owning, buying it from a man reconditioning old trucks. The lorry had been part of the Gilmore Dairy in Preston, Melbourne, one of the last dairies around to use horsedrawn vehicles. It was a four-wheeled, 8-by-5 feet flat tray, with a turntable at the front. Grae painted it crimson to finish it off.

He worked with 'Sammie' for three hours a day, two in the morning and one at night, learning how to get on with him and teaching himself how to operate the lorry. Later, he went to several courses taken by well-known Clydesdale trainer Noel Wiltshire, picking up the basics of safety: how to connect the traces to the horse correctly, getting the hooks on the harness to face the inside, keeping the harness neat, the horse's tail out of the way and the reins from being caught up in anything. He learnt the commands: 'Walk on, ponies'; 'Gee off' to turn right; 'Come near' to veer left; and 'Whoa'. He also learned about handling and horse care, the type of things any pony clubber would know from a young age.

A bond grew between horse and rider, as if they were looking after each other. Sometimes when they had finished their sessions Grae would sit in the paddock, thinking or relaxing, and Sammie would come over and nudge him on the ear then continue grazing, as if he were checking in to see that Grae was alright.

Samson once stumbled on a pile of wire hidden in long grass when they were out riding, falling like a 'tonne of bricks' with Grae coming down with him. Grae stood up, unhurt, but could see that Samson was caught. Worried that he might panic, try to get up and become more tangled in the wire that was already cutting into his legs, he took off his Drizabone and covered Samson's head with it. Samson lay there quietly until Grae had untangled the last strand of wire. It was complete trust.

Grae was going through an ugly divorce in the early 1990s, a time of angst and upheaval for him. Sammie became a comfort to him, a stable, calming presence. When he wasn't spending time with his children, Grae would escape the city to take solace in the quietude of Shannonvale. He let his hair grow long, grew a beard and built a ramshackle hut. He'd always felt uncomfortable in the city, he reflected, now he felt free. He imagined more and more living in the bush and taking his horse with him.

He lost his driver's licence for eighteen months for drink driving, in the city, and got around in the milk lorry, drawn by Sammie, doing gardening and odd jobs for people. He'd load his lawnmower and gardening tools onto its tray and drive from job to job. Samson would wait as he worked, harnessed with a dropped rein. Grae's customers thought it was marvellous—they'd tell him how their father used to do this or that with a Clydesdale or how they'd gone to school on a pony or similar, nostalgic at the thought of it. Sometimes when he'd finished work, Grae would drive the lorry down to North Ringwood, the next suburb along, to the Coach and Horses Hotel—just for an orange juice or a Coke, mind you, and a game of pool. He'd park the lorry where he could see it from a window so he could make sure Samson was alright.

In August 1993, Grae bought a second horse, a Clydesdale, to use for the hayrides. Rusty was a feathery-legged roan that looked the part. Grae looked the part too with his long beard, bushman's hat and vest. He put the finishing touches—three bales of hay—on the milk lorry, tested its rear lights and was all set to go with his plan to take rides along the Yarra River when the local council stepped in, worried that the lorry would damage the path along the river. Grae was incensed. The lorry had rubber tyres, not steel ones, how could that damage the gravel track? And the rides would add a tourist attraction to the riverbank, he argued. He complained to a local woman he knew who suggested talking to her sister Tess, who ran a café and had some experience in tourism matters. He asked Tess for the ammunition he needed to mount his case. Within four weeks, with Tess's help, Grae was plying the riverfront with Rusty plodding contently along from the Warrandyte Bridge to Stiggen's Reserve and back again. His customers loved the gentle ride, particularly the older ones who'd reminisce about Clydesdales.

But the dream ended one night two years later when Rusty ate a bucket of apples and developed a twisted bowel. The vet advised Grae to check on him the next day and, if he was still alive, take him to the Werribee veterinary clinic. Grae found Rusty dead the next morning; he'd thrashed about in pain and knocked his head on a tree.

By now Grae had started working with Tess in her café, closely. In 1996 they married. They arrived for the wedding at Stiggen's

Reserve in a carriage pulled by Samson. The couple moved to Shannonvale with Samson that year, Grae riding him along stock routes from Omeo to their property to save him the long, winding trip up in a float. The couple went on, over the years, to gradually build the remote area bed-and-breakfast known as Payne's Hut.

Samson took to the bush and the cattle droving Grae did with his neighbours with gusto—perhaps too much gusto sometimes as he shouldered the cattle into line with a little more push than the drovers liked! Grae and Samson helped move the neighbour's cattle to the Bogong High Plains for fifteen years. As ever, he would go wherever Grae pointed his head; whether it was up a steep incline or down a cliff, he never questioned it, full of trust. Once Grae rode him into a bog, following a cow and its calf. The cow and calf got through; Samson sunk up to his saddle flaps. As Grae clambered out of the thick mud, the horse sunk further. Grae urged him out—'Come on, boy!'—and with an almighty rear-end thrust, Samson exploded out of the bog like a cork. At other times Grae would let him have his head, galloping across cleared areas of the high plains, trusting that he would find his feet in the rough terrain.

In winter Grae worked at the ski fields of Mount Hotham to bring in extra money, rugging Samson for winter and checking on him when he came home at weekends.

He came home one winter at the end of the season to find Samson, by now 28 years old, not eating well. He took off his rug and as soon as he did the horse turned round and started biting the tendons of his back legs. Grae got him to walk—haltingly—and realised that he was in extreme pain. Who knows how long he'd been like that, he wondered.

He went inside to tell Tess what he was going to do.

He took the rifle and led his horse towards the slope at the rear of the property. Samson got halfway up the hill and stopped. 'Fair enough,' said Grae. 'If you don't want to go any further, I guess this is it.' He said his goodbyes and thank yous and cocked the rifle. Samson looked straight at him. Grae raised the rifle and shot him between the eyes. The aim was so accurate that Samson hit the ground without knowing a thing.

Grae had kept his pact.

GLOSSARY

baldy-faced horse with a very broad white blaze, sometimes over its eye(s)

barefoot trimming hoof care for unshod horses that mimics in trimming the way a hoof is naturally worn down in the wild

barrel race timed event in which a horse and rider complete a clover-leaf pattern around barrels

beast single cow of either sex

bending or pole-bending race timed event in which the horse and rider weave in and out of a line of vertical poles

breakaway string a loop of hay bale or similar used to tie reins or a lead rope of a horse to a fence so that if the horse pulls back strongly the bale will break

breastplate straps passing across a horse's chest to keep the front of the saddle or a harness from sliding back

bronc or bronco once used to refer to a feral or untrained horse, now more commonly describes a bucking horse used in rodeo events, often trained to buck

broodmare mare used for breeding

brumby a wild horse in Australia, usually descended from runaways

buckjumper horse that bucks, used for competitive sport

buckskin horse with a yellowy or dun coat with black mane and tail

campdrafting uniquely Australian sport in which the horse and rider work cattle

collected a horse is collected when it is 'on the bit', with its neck flexed at the poll and its body responsive and fully engaged

coloured horse horse with large patches of white and another colour, used to be called piebald (with black) or skewbald (white with any other colour)

conformation shape and proportions of a horse's body

crupper strap attached to the back of the saddle and looped under the horse's tail to stop the saddle or harness slipping forward

dilute a horse that's had the colour of its coat determined by a 'dilution gene', which causes the lightening of colour e.g. a chestnut horse with a diluted gene becomes palomino. Dilution genes include cream, dun, buckskin, pearl or silver genes

dogger animal to be used for pet meat or buyer of animals for this

draught horse generic term for a large horse from a range of breeds originally used for ploughing or hauling

Drizabone brand of full-length waterproof riding coat

entire male horse that hasn't been castrated

eventing equine competition with three types of riding: dressage, cross country and showjumping

flag race event in which the rider grabs a flag from the top of a peg (pole) furthest in a line of five pegs, returns and drops it in a bucket, then gets the next furthest flag and so on

forward a horse is said to be forward when it moves willingly with energy and impulsion

four-in-hand a team of four horses with reins joined together to allow the driver to control all of them (also six- and eight-in-hand)

frog V-shaped part on the underside of a horse's hoof that touches the ground and acts as a shock-absorber, important in circulating blood around the horse's legs

Garryowen Australia's premier show event for female riders over 18, held annually at the Royal Melbourne Show

girth band that passes around a horse's belly that holds the saddle in place

gooseneck covered trailer with a long neck that attaches to a hauling vehicle, can combine accommodation and horse floating

greenhorn new or inexperienced person

hack horse used for general riding (hacking) or can mean a worn-out horse for hire

hand or hands high measurement of horse height taken at the wither, equal to four inches or ten centimetres

hard-mouthed a horse that is unresponsive to the bit

harrows farm implement with spikes or upright discs drawn by a horse (or tractor) to break up the soil

haunch turn horse pivots on its hindquarters without moving forward

heavy horse another term for draught horse

hobble strap placed around a horse's legs to limit its movement

horn raised pommel common in Western saddles, used in cattle work to secure a roped animal

leg-yield horse flexes slightly from the poll away from the direction it was going

liberty work when a horse works without rider or tack, controlled by voice and body language

long-rein extended reins used to drive a horse from behind

lunging moving a horse around you in a circle on the end of a rope

mouth (verb) to teach a horse to accept and respond to a bit

nag old horse or one in poor condition

natural horsemanship a philosophy of training horses using techniques that work with the horse's natural instincts, ways of communicating and herd mentality, with an emphasis on training without force

neck-rein turning the horse by touching the reins on one side of its neck, causing it to turn from the pressure, used in one-handed Western riding

pacer horse bred for harness racing that moves in a two-beat lateral gait with its front and hind legs on one side moving together

palomino horse with a gold coloured coat with white mane and tail

para-equestrian form of equestrian sport practised by people with disabilities

passage dressage movement in which the horse moves with a highly elevated and energetic trot

passive trapping way of catching wild horses by luring them with salt, feed or water into a yard with a one-way gate that closes behind them

Percheron breed of draught horse originally from France with a black or grey coat

Piaffe dressage movement in which the horse is in a highly collected, elevated and cadenced trot on the spot

pigroot small buck

Pinto horse with large patches of white and any other colour

Pirouette manoeuvre where the hind legs stay on the same spot and the horse's front changes direction

poll part of the horse's head behind its ears

polo equine sport played by teams using long mallets with flexible handles to hit the ball through goal posts, like hockey on horses

polocrosse similar to polo but with elements of Lacrosse, using long-handled racquets; was developed in Australia

pommel raised front part of a saddle

pony (verb) to lead one horse from the back of another one that's being ridden
rein-back horse is guided by reins to move back in a straight line
reining Western competition in which the rider guides the horse through manoeuvres such as circles, spins and stops
remounts supply of fresh horses for the army
roll-back Western riding manoeuvre in which the horse comes to a standing stop and performs a 180-degree turn
saddle pad blanket on a horse's back underneath the saddle to cushion it and absorb sweat
shafts rigid bars extending from the front of a horse-drawn vehicle attached to the harness of the horse at its sides
shoulder-in lateral movement in which the horses bends its body so that its front legs track to the inside of the hind legs
side-pass horse moves both forelegs and hind legs directly sideways without stepping forward
snig to drag logs, branches or trees
snip small white marking between a horse's nostrils
Spanish walk movement in which the horse raises its forelegs high off the ground in an exaggerated motion
spell rest
standardbred breed of horse known for its ability to race in harness at a trot or pace. Distinguished from thoroughbreds by being shorter, longer and with a heavier bone structure
surcingle long strap that goes around the barrel of the horse to keep a saddle or other tack in place
tack any equipment used with horses, including saddle, harness, halter, reins
taffy dark brown horse with contrasting mane and tail, perhaps white or silver
traces side-straps, chains or lines that connect the harness and horse to the vehicle

transitions change of gait up or down, made in dressage at a given marker

trotter horse bred for harness racing that moves its legs in a diagonal gait

uveitis inflammation of part of the eye

Warmblood sport horse of medium weight and build created by crossing 'hot blood' breeds such as Arabs or thoroughbreds with 'cold blood' horses such as draught horses, though this has nothing to do with body temperature

Weanling foal that has been weaned but is not yet a yearling

Western riding American style of riding that evolved through cowboy work, controlling the horse with one hand while leaving the other hand free to rope, hold a gun etc.

Western saddle stock saddle originally designed to keep a cowboy secure and comfortable over long hours in rough terrain, characterised by a 'horn'

withers top of the horse's shoulder where the neck joins the body, between the shoulder bones

rein-back: [illegible] of gait up [illegible] down [illegible] given backward

trotter: horse bred for harness racing that trots at speed in a race

withers: top of the horse's shoulder; the highest point of the back between the shoulder bones

REFERENCES

Part I, Chapter 4

Outback Heritage Horse Association of Western Australia Inc
www.ohhawa.com.au

Part II, Chapters 6 and 7

For further reading about droving and chapter authors:
Horse Bells and Hobble Chains, Jeff Hill, 2003, Central Queensland University Press
The Privileged Few, Jeff Hill, 2008, self-published
More of the Privileged Few by Jeff and Cooee Hill, 2011, self-published

Part VI, Chapter 21

The Examiner 6 June 1979

Part VIII, Chapter 29

Chinese Anzacs: Australians of Chinese descent in the defence forces 1885-1919, Alastair Kennedy, 2012, O'Connor, A.C.T., self-published
Equitana www.equitana.com.au
Waler Horse Society of Australia www.walerhorse.com

Part X, Chapter 35

To find out more about Zelie Bullen's life:
Love Sweat and Tears, Zelie Bullen and Freda Marnie Nicholls, Allen & Unwin, 2013.

ACKNOWLEDGEMENTS

The interest in horses that propelled this book goes back years to the time I was a horse-mad girl pestering my parents for a pony. It worked, and thanks to them I enjoyed many weekends in my teenage years riding and pony clubbing. I am still grateful to my father for all the early mornings and days he spent floating Pepe the pony to gymkhanas and shows, and to my mother for taking us kids to riding schools when we were young and enduring Pepe eating all her camellias. I can only appreciate now the worry she must have felt finding me semi-conscious on the road and about to be taken to hospital by ambulance after a fall from Omar, my second horse. (I have ridden on treks from high country Victoria to beneath the High Atlas Mountains in Morocco since and make sure never to mention it to Mum too much until I get back in one piece!)

As a more recent horse owner I've been in contact with people who've helped me re-engage with horses, several of whom are in this book or who have friends or relatives in the book: Mark and Di McIntosh, Phil and Kris Ruby, and Jude Sadler are among them. Others who've helped with suggestions, contacts or photos include Thelma Churchill, Isabel Rhodes, Carolyn Ford, Geoff Bowen,

Karly Hickman at Equitana, Katherine Waddington, Julieann Martin and those at the Wonthaggi and Leongatha historical societies. Kylie Miller has been a great all-round support.

Heartfelt thanks to publisher Tracy O'Shaughnessy for commissioning the book and for her professionalism and warmth, and to others at Allen & Unwin, editor Laura Mitchell, and copy-editor and rider Susin Chow, for their enthusiasm and marvelous treatment of the manuscript.

Of course this book would be nothing without the wonderful people who've lent their stories to it, people who all share a love of horses. I've been touched by their tales and have learned much about the nature of these beautiful animals through the ones they've described.

To those people; I enjoyed every minute of talking to you. Thank you.